Composting
Utopia

I0810743

A Volume in the Series

ACTIVIST STUDIES OF SCIENCE AND TECHNOLOGY

Edited by

Sigrid Schmalzer

Composting Utopia

EXPERIMENTAL INFRASTRUCTURES FOR ORGANICS RECYCLING IN NEW YORK CITY

GUY SCHAFFER

University of Massachusetts Press
AMHERST AND BOSTON

Copyright © 2023 by University of Massachusetts Press
The electronic version has been made freely available under a Creative Commons (CC BY-NC-ND 4.0) license, which permits noncommercial use, distribution, and reproduction provided the author and University of Massachusetts Press are fully cited and no modifications or adaptations are made. Details of the license can be viewed at https://creativecommons.org/licenses/by-nc-nd/4.0/.

This book will be made open access within three years of publication thanks to Path to Open, a program developed in partnership between JSTOR, the American Council of Learned Societies (ACLS), University of Michigan Press, and The University of North Carolina Press to bring about equitable access and impact for the entire scholarly community, including authors, researchers, libraries, and university presses around the world.
Learn more at https://about.jstor.org/path-to-open/.

All rights reserved
Printed in the United States of America

ISBN 978-1-62534-769-5 (paper); 770-1 (hardcover)

Designed by Sally Nichols
Set in Scala Pro
Printed and bound by Books International, Inc.

Cover design by adam b. bohannon
Cover photo by Renée Peperone, *Food scraps at Know Waste Lands,* 2019.
Courtesy of the photographer.

Library of Congress Cataloging-in-Publication Data
A catalog record for this book is available from the Library of Congress.

British Library Cataloguing-in-Publication Data
A catalog record for this book is available from the British Library.

This work has been supported by the University of Massachusetts Press's Authors' Fund for Scholarship. For more information, visit
https://www.umasspress.com/authors-fund-for-scholarship/.

FOR EVERYONE

WHO

LOVES TRASH

Contents

LIST OF FIGURES ix
LIST OF ABBREVIATIONS xi
LIST OF COMPOST TERMS xiii
PREFACE xv

INTRODUCTION
Compost and Hope
1

CHAPTER ONE
A Genealogy of Municipal Organics Recycling
29

CHAPTER TWO
The NYC Curbside Organics Collection Pilot
BUILDING THE MARKET FOR ORGANICS RECYCLING
48

CHAPTER THREE
The NYC Compost Project
LOCALLY MEANINGFUL COMPOSTING
IN PARTNERSHIP WITH DSNY
75

CHAPTER FOUR
BK ROT
COMPOSTING NEOLIBERAL WASTE MANAGEMENT
TO MODEL JUST SUSTAINABILITY
117

CONCLUSION
Lessons from the Compost Heap
160

REFERENCES 179
INDEX 209

Figures

FIGURE 1. Charlie and volunteers watch as the toter tipper dumps out a toter on a new windrow. 86

FIGURE 2. Leah pilots a driving Jaylor full of mixed browns and greens at the Big Reuse facility under the Queensboro Bridge. 91

FIGURE 3. The Compost Promenade at Earth Matter. 98

FIGURE 4. Know Waste Lands. 119

FIGURE 5. The ritual at the Know Waste Lands open house. 128

FIGURE 6. Hilary's BK ROT comic. 137

FIGURE 7. The new bins on the concrete slab. 146

FIGURE 8. Victor poses with bike trailer at El Garden. 148

Abbreviations

AD	anaerobic digestion
BIC	Business Integrity Commission
C4B	Compost for Brooklyn
CORe	Centralized Organic Recycling equipment
CRAB	Citywide Recycling Advisory Board
DEP	Department of Environmental Protection
DSNY	New York City Department of Sanitation
EJ	environmental justice
KWL	Know Waste Lands
LESEC	Lower East Side Ecology Center
MSW	municipal solid waste
NRDC	Natural Resources Defense Council
STS	science and technology studies
SWAB	Solid Waste Advisory Board

Compost Terms

aerobic. Involving oxygen; this is the preferred state for most compost piles.
anaerobic. Without oxygen; most composters try to avoid anaerobic activity because it slows down a pile and produces odors.
anaerobic digestion. The decomposition of organic material without the presence of oxygen; this produces methane that can be captured.
browns. Carbon-rich, dry material for composting, such as woodchips or dry leaves.
compost. A soil amendment made from decomposed organic material.
composting. A controlled process of decomposing organic matter in the presence of oxygen to create a soil amendment.
compost tumbler. A contained compost system consisting of a free-moving compartment that allows users to mix and aerate compost.
feedstock. Anything going into a compost system.
greens. Nitrogen-rich, wet material for composting, such as food scraps or manure.
leachate. Liquid produced by a compost pile or landfill; can leak onto surrounding surfaces or into the ground.
microhauler. A small-scale waste hauler, typically operating locally and informally.
mixed solid waste. Unseparated waste that contains compostables, recyclables, and other materials.
organics. Material that was once alive and can be composted.
organics recycling. Any process that makes use of organic waste by breaking it down; generally includes composting and anaerobic digestion.
source separation. Sorting of wastes by waste generators.
static aeration. A system that forces air into a compost pile to increase efficiency.
tipping fee. A fee charged at a waste management facility for dumping off waste or recyclable material.
windrow. A long mound used for composting.

Preface

This book started in a dumpster behind a Trader Joe's in suburban Maryland in 2009. Or maybe on the way home from the dumpster, awkwardly sharing a back seat with stolen milk crates holding a greedy excess of bread, pies, muffins, seven-layer dip, cheese, and ill-advised sushi. At some point this became a weekly ritual, the drive between a cluster of punk houses in Northeast DC and a Trader Joe's just over the Maryland line. Impromptu trips with whoever happened to have a car. Over time, it transitioned from ritual to system. We kept a list of the houses involved in the dumpster share. We split the work so that we didn't double up on a dumpster run, and nobody had to go alone. One house kept the rescued food on display on a dining room table, an offering for guests. Another house stored juices in a freezer for leaner times. In a loose and uncoordinated way, we developed our own infrastructure for food recovery.

Driving home, windows down, eating garbage as we cruised through wealthy suburbs, it was so obvious to me that we could build our own world out of the scraps of the current one. I felt giddy and free. While the questions I ask in this book come out of science and technology studies and environmental sociology, the inspiration is back in that borrowed car, nestled among the milk crates.

I want to thank everyone involved in the process, from the dumpster to this book.

Thank you to everyone I shared dumpster food with in DC. You were always so excited to build new worlds and hang out in them.

Thank you to all my classmates and friends from graduate school, who made me feel excited to be part of an intellectual and political community of support and solidarity and creativity. This list is not exhaustive, or in any particular order, but thank you Gareth Edel, John Cluck, Ali Kenner, Kirk Jalbert, Sean Ferguson, Sonia Saheb, David Banks, Dan Lyles, James Wilcox,

Ben Brucato, Karin Patzke, Pedro De La Torre, Mike Lachney, Mike Bouchey, Lindsay Poirier, Laura Rabinow, Lindsay Karty, Dan Landers, Misha Wolf, and Allison Berkoy. Ellen Foster, thank you for late-night messy maps and crazy drives into the icy wilds of Vermont, for dance parties and swimming holes and the poetics of academic work. And Kate Tyrol, thank you for going to the dumpster with me, for speaking so honestly, and for always seeing through the intellectual knots other people made.

Thank you to my instructors and advisers, particularly the members of my dissertation committee. Kathy High, thank you for encouraging me to follow my love of decay, and thank you for all the time we spent researching fecal transplants, microbiomes, and human composting. Mike Fortun, thank you for helping me use queer theory to think about compost. Steve Breyman, thank you for never letting me lose sight of the goal of better waste systems. Abby Kinchy, I could have asked for no better adviser than you. You remained supportive, insightful, and patient throughout this whole process, from dissertation proposal to dissertation to book proposal to book manuscript to final copy. You have always helped me figure out the good parts in the bad things I write. You taught me how to do research. I will never be able to thank you enough for helping me survive grad school and teaching.

Thank you to the graduate students I've worked with. Maggie Mang, I've learned so much from your insightful care and your caring scholarship. Lee Nelson, thank you for being a brother in rot. And thank you to the undergraduate students who have talked through these ideas with me in classrooms. I have loved learning with you. Thank for being patient when I was too busy with revisions to offer feedback on your own amazing work. Thank you particularly to Evelyn Pae, for long discussions of Buddhist philosophy and its relevance to sustainability transitions.

Thank you to my colleagues at Rensselaer Polytechnic Institute. Jen Cardinal and Brandon Costelloe-Kuehn, you have been sources of joy, inspiration, strength, reflection, and care. Our conversations about experimental infrastructures, community work, and contingency resonate throughout this book.

Thank you to the generous editors and everyone else at University of Massachusetts Press. Sigrid Schmalzer, thank you for your many years of support and feedback over the time that it took for me to pull this book together. Thank you to my thoughtful reviewers for pushing the book in new directions. Your insights and suggestions made this book so much better than it originally was.

Thank you to the interdimensional queer theory book club: Hana Van Der Kolk, Jordan O'Jordan, and Louis Chavez. You were there with me when I fell in love with José Muñoz, Eve Sedgwick, and Ann Cvetkovich, and we worked together to figure out how to see our worlds through their words. Hana, thank you for your ongoing commitment to a queerer, more radiant, more awkward utopianism throughout our years of friendship.

Thank you to all of my community in Troy, for all of the experimental infrastructures we've built together. Thank you to everyone in little sangha, Cook Blub, Hudson Mohawk Magazine, Troy Zero Waste, Soul Café, garden partners, work sharers, and the mutual aid network.

Thank you to all of my friends in New York City who shared writing stress with me, particularly Joseph Hayden, Kyle Croft, Milo, Addison Vawters, Jeremy Sorese, Tait Mandler, Ben Haber, Rodrigo Chazaro, Jesús Maldonado, Colin Kennedy, MJ Knefel, and Gideon Oliver.

Thank you to all the composters who invited me into their projects, shared their thoughts with me, and provided feedback on this manuscript: Leah Retherford, Marisa DeDominicis, Charlie Bayrer, Louise Bruce, Marguerite Manela, Stephanos Koullias, Emily Dinan, and many others. Thank you particularly to everyone at BK ROT who made me feel like part of the team. Sandy Nurse and Renée Peperone, thank you for teaching me so much more than I thought I'd learn in my dissertation fieldwork. Ceci Pineda, thank you for your wisdom and generosity, and for recognizing that compost is obviously queer. Hilary Allison, thank you for letting me use your comic and for laughing with me among the worms. Victor Ibarra, thank you for keeping the system working all these years. Thank you to everyone who I've worked with on the board and on the team: Priya Mulgaonkar, Iyeshima Harris, Nora Tjossem, Dior St. Hillaire, Annel Hernandez, Dan Gross, Sonia Fountaine, Brys Peralta, Shaq Benn, Genea Foster, Felicity Slater, Kim Vallejo, and Fran Sanhueza. I tried out a lot of these ideas on you guys first.

Thank you to all my compost and community garden people beyond New York City who I've spoken with about community composting: Nora Goldstein, Brenda Platt, Jeffrey Neal, Kat Nigro, Tiffany Bess, Claudia Urdanivia, and many others.

Thank you, Samantha MacBride, for your ongoing insights and for talking through these ideas with me.

Thank you to the organizations whose generous support made this book possible: a Humanities, Arts, and Social Sciences Fellowship from RPI and a Doctoral Dissertation Research Improvement Grant from the National Science Foundation.

Thank you to my family. Nora and Kristen, thank you for always thinking more deeply and reflexively than I could. Thank you for sharing references, asking questions, and inviting me in to your own work. Mom and Dad, thank you for talking me through every twist and turn. John and Elaine, thank you for your love and rebelliousness.

Most of all, I want to thank Kevin and Nova. Kevin, thank you for being proud of me for writing a book this whole time, even when I was pretty sure I couldn't write a book. Thank you for circumnavigating Shaver Pond when we needed to, for talking through research dilemmas, for helping me figure out what I wanted to say, for dealing with my gross rotting food scraps and smelly compost piles, for watering the garden, for being a beacon of light, for creating beauty in our lives, for imagining futures with me.

And Nova, thank you for reminding me when it's time to eat and go outside, for being excited when I'm done writing, for waking me up in the morning.

Composting Utopia

INTRODUCTION

Compost and Hope

This book is about compost and hope, and the way that both are produced in experimental infrastructures in New York City. The infrastructures at the emotional core of this book—community compost sites and compost microhaulers—take in organic waste and transform it into compost that feeds street trees, mitigates flooding, grows food, and supports neighborhood gardens. These are grassroots experiments in sociotechnical systems building, where activists articulate alternative networks of relationships between technologies, values, knowledges, microbes, information systems, habits, relationships, wastes, and people. In the process, they create spaces in which people can feel active hope for a sustainable, just future based on the visions and values of on-the-ground communities. Experimental infrastructures for compost aren't solutions to the city's waste problems, but they have helped me to feel hope for the future of waste management. I hope that by reading about them, you might feel the same.

Grassroots infrastructures are a common response to the perceived inadequacy of dominant systems. Social movements often construct their own informal grassroots systems to enable new and better ways of connecting the world. The architects of feminist hackerspaces, community wireless networks, food coops, bike rescues, free clinics, neighborhood gardens, community recycling centers, underground abortion clinics, needle exchange programs, lesbian separatist communes, and cooperative houses, along with countless other projects, have imagined and constructed alternatives to the systems at hand (Fox, Ulgado, and Rosner 2015; Forlano 2016; Zitcer 2015; Bradley 2018; Nelson 2011; Martinez 2010; Lounsbury, Ventresca, and

Hirsch 2003; Bart 1987; McLean 2011; Sandilands 2002; Barrett 2013). Many of these projects are made necessary by conditions in which people are unable to access the resources that are needed for flourishing. Grassroots infrastructures can even help to maintain these conditions: they might distract public concern, legitimate existing structures, or reinforce dominant understandings of social problems (MacBride 2011; Eliasoph 2013; R. Keil 2009). But in many of these projects, activists go beyond trying to fix problems in the here and now, and articulate new sociotechnical imaginaries: new visions of how people can use technology to build a better world (Jasanoff and Kim 2009). They demonstrate the possibility of a future that escapes dominant narratives about what progress looks like.

I'm using the term *experimental infrastructures* to describe systems that, intentionally or unintentionally, point toward alternative ways of organizing the world. For example, BK ROT, a compost microhauler I examine in this book, enacts a system in which wastes are recycled locally, the work of environmental stewardship is valued, and young people of color and immigrants have agency in urban design. Community composters and microhaulers turn activist values and local visions into inhabitable infrastructures, modeling new approaches to urban waste. Experimental infrastructures have the potential to function as *concrete utopias* in the sense described by Ernst Bloch: present experiences that orient people toward counterhegemonic futures (Levitas 1990b).

New York City's experimental infrastructures for compost have worked in complex relationship with one of the largest waste systems in the world: the New York City Department of Sanitation (DSNY). During the period when I was conducting fieldwork for this book, DSNY was also oriented toward a future in which the city had a healthier relationship with waste, as expressed in its goal of zero waste by 2030 (abbreviated as "0X30," which was printed on the side of the city's dual-bin collection trucks). As a part of this effort, New York City conducted a municipal curbside organics collection pilot program that diverted food waste from households and institutions for composting and anaerobic digestion. Community composters and microhaulers played a role in the city's organics collection pilot program: they trained residents to separate food scraps, demonstrated the power of compost, educated New Yorkers about the importance of waste diversion, and advocated for change in the city's waste systems. DSNY has at various times learned from community composters and microhaulers, partnered with them, restricted their activities, and threatened their existence. The growth of the city's curbside organics collection stalled out in 2019, and the program was put on pause in 2020, for the stated reason of budget cuts in response to the COVID-19 pandemic. Many of the city's grassroots compost infrastructures have outlived municipal curbside pickup—at least for now.

This book describes three infrastructures for organics recycling in New York City: municipal curbside collection as coordinated by the Department of Sanitation, community composting as organized through the NYC Compost Project, and the microhauler BK ROT. All of these infrastructures address the excess organic material in New York City's waste stream. But community composters and microhaulers are doing this in ways that are locally meaningful, addressing the needs and interests of their communities: green space, good jobs, opportunities for young people, local resource loops, and healthy soil. In contrast, the curbside organics pilot, while it more directly tackled the massive waste stream of New York City, has been relatively limited in terms of the ways it can fit into New Yorkers' varied understandings of the sustainable city.

In telling this story, I draw on the interdisciplinary field of science and technology studies (STS) to describe how values are built into waste infrastructures and how grassroots composters politicize the knowledge and technology of organics recycling. In experimental infrastructures, organizers demonstrate alternative sociotechnical possibilities that are rooted in local values and respond to the complexity of local contexts. Experimental infrastructures function within existing structures and are shaped by them; they do not exist outside neoliberalism, racial capitalism, colonialism, or other structures of domination. This can limit their possibilities, but it also means they offer alternatives that respond to the present. I use the metaphor of *composting*, borrowed from feminist STS, to describe how grassroots composters repurpose their contexts to build something new.

Experimental infrastructures for composting redefine sustainability, envision new futures, and assemble new communities that can work toward these utopias. Most importantly, they kindle a specific experience of hope: not of promise, but of possibility—that sustainability could be imagined otherwise, starting where we are.

WASTE HANDLING IN NEW YORK CITY: A SUSTAINABILITY TRANSITION?

In 2020, the New York City Department of Sanitation collected 4.7 million tons of discards from households and institutions (DSNY 2021). Of this material, 17% was recycled, 1% was composted, and 82% was shipped to landfills and incinerators across New York State, New Jersey, Connecticut, Pennsylvania, Ohio, Virginia, South Carolina, and Kentucky (DSNY 2021; Murphy 2015). Commercial haulers move another three million tons from businesses each year (Stringer 2018), along with 6.4 million tons of construction and demolition waste (Goldenberg and Muoio Dunn 2020).

For most users of New York City's waste system, this Herculean feat is relatively humdrum. City dwellers can witness, without much surprise, the ordinary miracle in which mountains of trash bags disappear from their sidewalks overnight. This is part of the magic of infrastructure: when working properly, infrastructures blend into the backdrop of everyday life for many users. Susan Leigh Star describes working infrastructures as "by definition invisible, part of the background for other kinds of work" (1999, 380).

The invisibility of infrastructures hides the ways they shape the world, defining what people can do, what people want to do, and how people think (Star 1999). Infrastructures for waste management make city life possible: they keep streets clear, prevent the spread of disease, and beautify public space (Melosi 2005; Strasser 1999). They also shape the ways that city dwellers understand our relationships with our environments and with consumption. Gay Hawkins describes how tidy systems for waste management make it possible for people to imagine our household disposal habits as self-contained and separate from the world around us (2006). The disappearance of garbage through collection, landfilling, incineration, and recycling is a necessary complement to consumerism—not only for the practical purposes of disposal but also to hide the wastefulness of consumption from public consciousness (Strasser 1999).

If you have spent time in New York City, you might recognize that DSNY's disappearing act is not entirely convincing. DSNY might be able to make your personal waste vanish, but you'll still get stuck behind a collection truck on a narrow street, or smell the steaming trash piles on a hot summer night, or watch the video of "pizza rat" carrying a discarded pizza crust through a subway station. In an episode of the sitcom *Broad City*, Abbi Jacobson yells at the customers in a New York City copy shop, calling them "garbage people living on garbage island" (Jacobson and Glazer 2014). Garbage is central to the city's identity, despite DSNY's efforts to make it disappear. Brian Larkin suggests that invisibility is only one mode of perceiving a given infrastructure. At other times, from other vantage points, the presence of infrastructure is overwhelming. Instead, Larkin examines the "poetics" of infrastructures: the ways that infrastructures are used to define structures of meaning in everyday life (2013). The infrastructures that make waste invisible for some people make waste hypervisible for others.

The city's waste system is very present for the people who live with material transfer facilities, truck routes, and recycling facilities. In 2018, three-quarters of New York City's solid waste was processed in four community districts (out of 59) in North Brooklyn, the South Bronx, and Southeast Queens, all of which are predominantly Black or Latinx and low-income (Neuman 2018). This has negative impacts on the health and well-being of these communities, caused by pollution from transfer facilities, the concentration of hauling routes and the particulate matter in diesel truck exhaust, and the clear message of turning a handful of neighborhoods into the city's trash dumps (Wood 2019). New York City is not unique in this regard; waste infrastructure is historically wrapped up in the production of environmental injustice. Landfills, incinerators, radioactive waste facilities, transfer stations, and materials recovery facilities have historically been placed in Black, Latinx, and working-class communities, on Indigenous land, and in developing countries, shouldering residents with the burden of toxic materials, unknowable risks, nauseating smells, and social stigma (Pellow 2003; Bullard 1990; Akese and Little 2018; Jantz 2017; Foote and Mazzolini 2012). Recent work in critical environmental justice studies has argued that racism has shaped not just the siting of waste facilities but also the design of waste-generating systems, because the production of wastes depends on the availability of specific places—primarily places inhabited by people of color—as "sacrifice zones" for managing and abandoning wastes (Pulido 2018; Dillon 2014). The overproduction of wastes depends on the white supremacist, colonial idea that the homes and bodies of people of color can be polluted in order to create clean environments for white people (Liboiron 2021; Pulido 2018).

The waste system is likewise hypervisible for the tens of thousands of waste workers who make the public, private, and nonprofit waste sectors of New York City function. This is dangerous work: waste collectors are three times as likely to die on the job as police officers or firefighters (Bureau of Labor Statistics 2021; Feldman 2018; Royte 2006). This is particularly true in the private waste industry, where fatality and injury rates are significantly higher: nationwide, 85% of waste industry fatalities occur in the private sector. In New York City, the private sector also tends to offer lower pay for more precarious work, and has a predominantly Black and Latinx workforce, while DSNY is predominantly white (Newman 2016). In both the public and private sectors, waste labor is unpleasant, stigmatized, and often thankless. Robin Nagle, resident anthropologist at DSNY, describes the invisibility of waste labor: "sanitation workers are *willfully* unseen by the public" (2013, 23).

Environmentalists, community organizers, policymakers, industrial actors, entrepreneurs, and others have come to see New York City's waste system as problematic, and have worked to change the way the city disposes of waste. Campaigns have pushed to end incineration (Barbanel 1985; Showstack 1997), to close down the Fresh Kills Landfill on Staten Island (Melosi 2016), to redistribute waste infrastructure (Casey and Mehrota 2011), to use cleaner trucks for disposal (Sze 2006), and to end the city's recycling program (Tierney 1996). These campaigns have achieved varying levels of success in reshaping the city's waste system.

In 2013, responding to environmentalists, environmental justice advocates, policymakers, and industry, DSNY started exploring ways to divert organic material from the waste stream and toward composting and anaerobic digestion. New York City's turn toward organics recycling was part of a nationwide wave: in the United States, between 2005 and 2017, the number of communities with composting programs increased over thirteenfold, from 24 to 326 (Streeter and Platt 2017).

Scholars and policymakers use the term *sustainability transitions* to describe processes through which existing infrastructures are replaced with more sustainable ones. The multilevel perspective describes sustainability transitions as the result of interactions between *niches*—protected spaces for innovation—and *sociotechnical regimes*—obdurate, established systems that consist of dominant technologies, institutions, and values (Geels 2005; Loorbach, Frantzeskaki, and Avelino 2017). Changes in the *sociotechnical landscape*—the larger political and cultural context—exert pressure on regimes and niches, causing shifts in regime activities or even the substitution of regime technologies by niche innovations (Geels 2011).

The multilevel perspective can frame a useful history of New York City's fledgling organics recycling program: since 1991 (when recycling was introduced), the city has had a fairly well-defined regime for managing waste and recyclables, made up of the interlocking relationships among DSNY, commercial haulers, waste disposal habits, attitudes toward recycling, interstate garbage routes, and many other pieces. In the 2010s, landscape pressures—including EPA policy, local and national movements for waste reduction and environmental justice, increased tipping costs, and so on—created space for the city to pilot curbside organics collection. Large-scale compost facilities and curbside collection had been developed in other cities in the United States, and large firms such as Waste Management had already invested in this infrastructure, so these technologies were available for incorporation into New York City's system. The city government undertook initiatives to incorporate these new technologies: new laws requiring organics diversion, new offices and mandates within DSNY, and new contracts with long-distance haulers and processing facilities. In 2013, DSNY started an organics collection pilot program in three neighborhoods, with plans to expand the pilots over time. While these changes were significant, they did not demand substantial restructuring of the incumbent regime, and involved many of the same elements as existing waste and recycling systems, such as public-private partnerships with large waste firms, material transfer facilities, and interstate hauling of waste. The emergent system was only partially incorporated into the regime, challenged by low participation rates and the financial shock of the COVID-19 pandemic.

Sustainability transition research has traditionally focused on large-scale infrastructural changes, centering the work of large firms, policymakers, and designers who lead the construction of municipal or national infrastructures that are needed to build sustainable worlds (Köhler et al. 2019). The multilevel perspective treats niches as important insofar as niche innovations have the capacity for wider adoption. These approaches fail to capture the coordination across scales that happens in sustainability transitions. In recent years, there has been a turn toward conceptualizing sustainability transitions as emergent phenomena involving both the top-down processes of policy and infrastructure design, and the bottom-up processes of shifting household practices, social movement action, and grassroots innovation (Belda-Miquel, Pellicer-Sifres, and Boni 2020; Hargreaves et al. 2011; Shove and Walker 2010). Not only does the traditional top-down perspective make it difficult to see bottom-up processes, it prioritizes the problem framings and solutions that are preferred by powerful actors. Adrian Smith suggests that the top-down perspective runs the risk of "construing civil society as a problematic medium that has to be managed or pushed aside" (2012, 183). Describing transitions from "below" offers the potential to include the agency, ideas, and values of those who are excluded from formal planning processes.

Social movements can shape sustainability transitions by shifting the cultural landscape and by changing the ways individuals relate to their infrastructures (A. Smith 2012). In a context where sustainability has become depoliticized (Swyngedouw 2009), social movements and civil society play a vital role in challenging existing discourses and articulating new ideas that can shape environmental policy (Greenberg 2013; Desfor and Keil 2004). Civil society can also mobilize political support for experimental systems (Amenta 2006). Zero waste and environmental justice movements have influenced New York City waste systems through reports, campaigns, testimony at public hearings, the movement of activists into DSNY and city council, and many other methods.

Civil society can also develop new social and technological innovations that shape sustainability transitions. Sanitary reformers, organic gardeners, recycling advocates, and others have pioneered some of the techniques and frameworks that were brought together in the construction of the city's organics recycling pilots (Melosi 2005; Lounsbury, Ventresca and Hirsch 2003; Blum 1992). Gill Seyfang and Adrian Smith coined the term *grassroots innovation* to describe "networks of activists and organisations generating novel bottom–up solutions for sustainable development; solutions that respond to the local situation and the interests and values of the communities involved" (2007, 585). While mainstream innovation processes are generally motivated by profit, grassroots innovators are more likely to be motivated by "social need and ideology" (591). The difference in motivation means that grassroots innovations often frame social and environmental problems differently from the regimes that they challenge or complement, and are able to come up with novel solutions.

Grassroots composting initiatives are an important part of the city's "transformation from below" (Smith et al. 2015). They train their users on the techniques of source separation, and help to make sense of composting. They innovate within the system, and consult with DSNY on solving technical challenges. Community composters are already celebrated by DSNY; its 2015 Organics Collection report explains that the residential organics collection pilots "built on decades of hard work among community composters dedicated to education and local sustainability" (20).

But the bottom-up work of grassroots composting goes beyond meeting and supporting the top-down work of municipal collection. These systems embody an alternative politics of waste management, often rooted in social movement values and local priorities. Many embrace the values of "just sustainability," centering both environmental and social concerns in their design process (Agyeman, Bullard, and Evans 2002). Community composters and microhaulers might prioritize local resource cycles, healthy soil, good green jobs, community activities, local food, or other values.

As hundreds of US cities introduce municipal organics collection, the majority of these programs have reproduced the structures of existing waste and recycling systems. They work with existing infrastructures and institutions, crafting compost systems that prioritize diversion rates and cost effectiveness. While the institutionalization of organics recycling is promising for waste reduction, it also involves the standardization of a specific and limiting approach to organic waste that has been defined by policymakers and industry. The evolution of this approach is the focus of chapter 1.

Community composters and microhaulers support the transition to organics recycling, but often challenge the dominant framings of urban organics recycling. Standard transition narratives do not include projects like these, which don't aspire to replace incumbent systems. But their challenge from below is an important part of New York City's organic waste transition, and centering their stories provides an alternative perspective on the politics and possibilities of sustainability transitions.

ENVISIONING ALTERNATIVES WITHIN NEOLIBERAL WASTE MANAGEMENT

It's easy for stories about grassroots innovation to slip into the "dystopian events repackaged as feel-good story" genre,[1] in which accounts of the hard work and creativity of people struggling to survive can normalize the forces that made survival improbable in the first place. The interactions between New York City's organics recycling regime and grassroots composters need to be understood in the context of neoliberal waste management, which has shaped the conditions in which grassroots projects take shape and find support.

I'm using the term *neoliberalism* to refer to a set of social, cultural, and policy shifts that began in the 1980s in the United States, in which free-market ideology has come to dominate as a way of understanding society, politics, the

1 This genre has been named and many examples have been archived by a Facebook group of the same name ("Dystopian Events Repackaged as Feel-Good Stories," 2019).

environment, education, the family, and other domains of life. This transition has replaced many public amenities with marketplaces or public-private partnerships (Harvey 2007; Miraftab 2004). Services such as the post office, public transportation, national parks, or education that were once understood as public goods have been restructured as profit-seeking enterprises (Connell, Fawcett, and Meagher 2009). Marketization also brings with it a transformation of subjectivity: people's identities within institutions—as citizens, students, artists, partners, and so on—are translated into the economic terms of entrepreneurs and customers (W. Brown 2017; Foucault 2010; Pollack and Rossiter 2010; Maniates 2001). Fundamentally, neoliberalism can be understood as the extrapolation of capitalist ideology into a theory of everything.

By *neoliberal waste management*, I mean a structure and understanding of waste management shaped by what Mark Fisher describes as "capitalist realism"—a stance in which markets and transactions become universal descriptors of human activity, and it's impossible to imagine a world outside of capitalism (2009). The impact of neoliberal ideology can be seen in the structure of the waste industry, the dominant policy approaches, the culture of consumers, the ways activists respond to waste, and the (often very limited) ways designers imagine alternatives to the waste system. Neoliberal waste management is depoliticized, or "postpolitical," and economic rationality is more valued than human values or democratic design processes.

Neoliberal waste management privileges markets but does not imply a total lack of public infrastructure. There are nonmarket elements present throughout waste systems—for instance, many US cities (including New York City) have publicly operated residential collection programs and privately owned commercial collection (Feldman 2018). There are also many practices that critically challenge market logics, such as the giveaways for compost and leaf mulch in many US cities, which redistribute organic waste through nonmarket transactions. Following J. K. Gibson-Graham, I describe the ways neoliberalism has shaped the industry and our imaginations around waste management, without painting over alternative economic structures that people have worked hard to build (2006). I will highlight three specific ways that neoliberal ideology manifests in waste management, and what this means for the people who live with waste infrastructure.

First, neoliberalism encourages states and enterprises to approach waste as a market opportunity. In policy, this has meant the rollback of public waste services and the creation of markets for waste management. Beginning in the 1980s, the withdrawal of federal support for municipal waste services led increasing numbers of cities to contract with private waste industry for collection, hauling, and landfilling (Melosi 2008). Additionally, a series of US Supreme Court rulings spanning from 1978 to 1994 declared it unconstitutional for states and counties to restrict the movement of waste, which enabled waste corporations to expand operations to cover larger and larger swaths of the country (O'Leary and Walsh 1995; McCarthy 2003), and to consolidate the industry through mergers and acquisitions (Melosi 2008). The profit motive for private waste management firms encourages them to cut costs, which in many cases has led them to be worse contributors to environmental justice than municipal waste systems: they site more waste facilities in communities of color and working-class communities (Wood 2019), and often hire workers for more dangerous, more contingent, and lower-paying jobs (Newman 2016; Pellow 2003). In New York City, the lack of coordination among private firms has also led them to create unnecessarily redundant infrastructure that particularly burdens communities of color (Wood 2019).

In the neoliberal imagination, the best way to address the problems of waste systems is to manipulate the market: either by increasing the cost of disposal, through pay-as-you-throw (PAYT) policies and tipping fees, or by increasing the value of wastes, through the creation of markets for recyclables. In many cases, increased disposal fees can reduce wastes (Skumatz 2008; Karidis 2016), but even when they work to reduce wastes, their impacts can be distributed unevenly. Increasing the cost of disposal can punish those who don't have access to recycling options (Skumatz 2008), or encourage people to dump waste covertly (Botetzagias et al. 2020). Policies that reframe wastes as resources in new recycling markets can help to legitimize the continued use of marginalized places as dumping grounds without significantly changing the relationships between people and waste (Lucier and Gareau 2015; Alexander and Reno 2012; MacBride 2011; Pellow 2003).

Fundamentally, market-based solutions to waste problems are often ineffective because they treat wastes as a given, and fail to tackle the systems of production and consumption that produced these wastes in the first place (Hird et al. 2014). Samantha MacBride's work on recycling in New York City describes how the city has made heavy use of markets and market-based policies for collection, hauling, and processing, including contracts with private recycling firms, bottle bills, and the encouragement of small-scale reuse enterprises (2011). MacBride finds that these tools not only failed to reduce the city's waste stream but also diverted attention away from the industrial actors who are most responsible for waste, and onto individual consumers.

This is a second way that neoliberalism shapes waste management: consumer behavior is understood as the primary cause of waste, and as the most appropriate site for intervention. People who discard waste are often framed as *waste generators* who are responsible for creating waste and recyclables and, therefore, responsible for properly disposing of them. In this framework, waste is a question of individual morality, to be solved through education and incentives (MacBride 2011). This idea is so matter-of-fact within the current waste management ideology of the United States that it's hard to imagine anyone *but* consumers as responsible for generating waste! But contrast this with an extended producer responsibility model, for example, in which producers are responsible for goods and packaging at the end of a product's life, and are therefore encouraged to design goods that last, that are repairable, or that come in reusable packaging (Walls 2006). In a context where consumer products are excessively packaged, repair is often discouraged, and zero-waste products are sold at a premium, framing consumers as a source of waste is, at best, an incomplete picture.[2]

Finally, volunteer labor plays an important part in neoliberal waste management. The withdrawal of state support for basic services, coupled with the celebration of individual responsibility, means that neoliberal systems often rely on organized volunteering to meet the needs that are not met by market forces. Nina Eliasoph explains that such volunteer work often feels productive but distracts from the systemic problems that caused those needs in the first place (2013). Samantha MacBride describes these positive feelings as *busy-ness*: "a fulfilling sense of work and achievement that often brings positive side effects but fails to reach the central effect" (2011, 6). MacBride identifies busy-ness in household recyclables sorting, which feels productive, but often falls short of its eco-friendly image. We can also feel busy-ness in large-scale volunteer events such as litter pickups and food rescues. In small ways, these projects address the immediate problems

2 Of course, consumer behavior remains an important part of any waste system, and sustainability transitions rely on the work of consumers in the construction of new infrastructures. But when consumers are the focus of waste policy—as in PAYT systems, bottle bills, or education campaigns—it can become difficult to conceptualize alternative approaches to waste (Maniates 2001). Understanding waste as a problem of individual morality does not encourage designers, policymakers, or activists to rethink system-level approaches to production and distribution (Hird et al. 2014).

of litter, or food waste and food insecurity, but these problems could be more effectively addressed by reducing the use of disposable materials or by designing a food system that prioritizes feeding people over corporate profit. The good feelings that accompany volunteering make it seem like an important response to social problems rather than part of the structure that continues to reproduce those problems.

To sum up, neoliberal waste management is characterized by the use of markets—both for waste services and for recyclable materials—in which consumers are understood as the source of waste problems. Solutions to waste problems focus on individual action—education campaigns to encourage better consumption, or volunteering through the appropriate channels. Despite the continued central role played by DSNY in organics collection (which was offered free of charge to all residents in the pilot zones), the neoliberal influence on New York City's organics recycling is present. It can be seen in the creation of markets for commercial organics recycling; the use of public-private partnerships for transfer stations, haulers, and processors of organics waste; the targeting of individual consumers as the cause of excess waste; the work of large, volunteer-powered food rescue organizations; and regulations that encourage community composting as an act of environmental goodwill, rather than as vital labor that should be compensated. These policies have created space for large waste industry firms to take on new contracts with the city, and to collect, haul, and process organic materials in a way that largely reproduces the dynamics of existing systems for waste and recycling.

At the same time, some grassroots projects have been able to flourish within this context, embodying new and sometimes radical approaches to the city's waste problems, and making creative use of the cultural and regulatory frames set forward by the city. A handful of microhaulers have been able to take on the new market for collection that was created by the commercial organics ban. These small-scale haulers embody a variety of different ideas about how the city should manage waste, such as prioritizing good jobs, local resource cycles, or youth empowerment. While individualization can limit people's imagination for institutional change, the emphasis on individual responsibility may also contribute to the demand for residential organics collection outside of the city's pilot zones, which has helped to build the market for microhaulers and encouraged people to get involved in other community compost projects. Many "waste generators" make use of these services because they want to "feel better about their trash." One entrepreneur, discussing his dream for a bike-powered hauling project, quipped that he had an opportunity to "monetize white guilt" in this new market. Volunteer labor powers or supports nearly all of the city's grassroots compost systems.

Neoliberal waste management has enabled two different but related approaches to organics recycling: a formal system that includes large corporations and public-private partnerships, and many informal systems that include small-scale, local interventions and grassroots innovation. While both have been able to operate side by side in the regulatory, cultural, and infrastructural landscape of New York City, the formal system is reinforced through contracts with the city, legal structures, and the large size of the relevant departments and corporations. The informal systems have much more tenuous relationships with city agencies, often occupy legal gray areas, and frequently work on a shoestring budget.[3]

In some ways, the relationship between the formal organics recycling regime and the informal niche-based innovations looks like a David-and-Goliath story, where small but nimble grassroots composters struggle to find a foothold in the large and inflexible municipal organics system. But the relationships between formal and informal systems are not simply competitive; Ananya Roy describes formal and informal systems as interdependent elements of neoliberal development (2009). Roy argues that informal systems make neoliberal development possible by meeting the needs neglected by formal systems. Anna Tsing's concept of "salvage accumulation" is helpful for framing this relationship: Tsing describes salvage as a "process through which lead firms amass capital without controlling the conditions under which commodities are produced" (2015, 63). When capitalist production fails to create the conditions that can support life, informal systems often, but not always, pick up the slack to supply the laborers, skills, places, and materials that are necessary for capitalist production. In return, formal systems support informal systems insofar as they create spaces in which they are relevant, and allow them to continue to exist (Roy 2005; Roy and AlSayyad 2004; Miraftab 2004). In New York City, informal composting systems meet many residents' needs for sustainable disposal options,

3 The distinction between formal and informal infrastructure comes from studies in the global South, which have used the term *informality* to describe a bottom-up, decentralized, illegal or quasilegal kind of city building, exemplified by slums, contingent labor, and unregulated economies (Elsheshtawy 2011; Roy 2005). Scholars have also examined informality as an important part of cities in the global North, where garage sales, food trucks, and unpermitted housing operate at the edges of legality (Mukhija and Loukaitou-Sideris, 2014). Informal urbanism has been studied as both a product of neoliberal retrenchment and uneven development that is only barely able to meet people's basic needs, and also a site of ingenuity and resilience, where people are able to live outside of oppressive institutions (Bailey et al. 2018; Chien 2018; Amoako 2018).

green spaces, and healthy soil—needs that the city's formal waste, planning, and food systems have failed to meet. They also help to nurture skills and attitudes that can be used to support the organics collection pilots. In return, DSNY and other agencies have enabled spaces in which many of these systems have been able to keep functioning, at least for now.

While formal and informal compost systems depend on each other, formal systems have the upper hand: formal processes "determine which forms of informality will thrive and which will disappear" (Roy 2005, 149). Community composters and microhaulers expressed fear that the formal institutions of waste management, the Parks Department, or housing development would take over their physical spaces and market niches, making it impossible for them to survive. In 2015, the Department of Housing Preservation and Development earmarked 43 community garden sites for affordable housing development, including several community compost sites. While protest from community gardeners convinced the department to transfer 34 of these sites to the Parks Department, 7 active gardens were demolished, and the other gardens had to spend a year fighting for their right to exist (Nir 2016; Nonko 2016). Community composters often have only informal claims to the spaces on which they operate, and it's not difficult for formal institutions to negate these claims when needed.

Formal systems can also subsume informal systems. This may provide stability but also exclude ideas, people, and practices that are not consistent with the values of formal systems (Joshua Goldstein 2021; Sternberg 2013). Some community compost systems have been formally incorporated into DSNY through the NYC Compost Project, which has supported these systems while also limiting some of their activities. A program launched in 2016 by the Business Integrity Commission (BIC) allowed nonprofit microhaulers to haul food waste from businesses, which had previously been illegal, in the interest of collecting data to figure out how best to regulate microhaulers. However, this left for-profit microhaulers struggling to figure out how they fit into the new economy of small-scale composting.[4]

The grassroots composting systems that I describe in this book were built

4 Formalizing informal waste management often creates precarious legal and economic gray areas that limit the possibilities for waste laborers. For instance, Carolina Ana Sternberg describes how informal trash pickers in Buenos Aires—*cartoneros*—were formalized as the city underwent a variety of neoliberal reforms in the early 2000s (2013). By building resource recovery centers and registration systems, providing uniforms to *cartoneros*, and rebranding them as *recolectores urbanos* (urban gatherers), the city turned its scavenger class into a uniformed sanitation force without having to provide formal employment or benefits. Similarly, Joshua Goldstein describes how informal recyclers in China have been excluded from the formations of modern citizenship (2021).

in response to neoliberal waste management and other capitalist processes of city building: gentrification, divestment, food apartheid, and others. Grassroots composters do their best to reduce waste, connect people with their neighbors, support community gardening, center the voices of youth, and build a more livable city. They do all this in a context where sustainability, justice, and human survival often take a back seat to market creation. But informal systems aren't separate from the ideologies and institutions of neoliberal waste management—they are enabled by the same systems and support them in some ways.[5] Storytelling about experimental infrastructures needs to account for both their embeddedness in their neoliberal context, and their potential to imagine alternatives to market rationalities.

FINDING UTOPIA IN THE COMPOST BIN

A central argument of this book is that many experimental infrastructures function as concrete utopias, encouraging users and maintainers to imagine and work toward new sociotechnical futures. The term *concrete utopia* comes from Ernst Bloch,[6] who differentiates concrete from "abstract" utopias that distract from the present with the promise of a better future. Concrete utopias might still be unreachable but reorient human activity toward an alternative vision of the future. Ruth Levitas explains that concrete utopia is "actively bound up in the process of the world's becoming, as an anticipation of the future . . . and, through its effects on human purpose and action, as a catalyst for the future" (1990b, 87). Concrete utopias function to "educate hope" (1990a), bringing together hope for a better future with an analysis of the problems of the present and the possible avenues for change.

This is complementary to the claim that alternative technologies can have an impact on larger sociotechnical systems through co-optation, diffusion, and incorporation into sociotechnical regimes, which has been demonstrated in the literature on sociotechnical transitions, social innovation, and alternative pathways in technology (Hess 2007; A. Smith 2012; Geels

5 Roger Keil characterizes this as "roll-with-it" neoliberalization: a moment in which "political and economic actors have increasingly lost a sense of externality, of alternatives (good or bad) and have mostly accepted the 'governmentality' of the neoliberal formation as the basis for their action" (2009, 232). It is not straightforward to piece apart the processes of contesting and supporting neoliberal waste management.

6 Bloch's work has been generative for queer theory, performance studies, economic geography, and other fields. Outside of this genealogy, though, thinkers in many fields have described the generativity of utopian visions, examining how designers, organizers, artists and writers, build new ways of living, thinking, and feeling (Grandoit 2020; Cruz 2020; Wright 2010; Monani 2009; Delany 2012).

2011). Experimental infrastructures are not simply prototypes that might or might not get adopted into larger systems, they are spaces that have "effects on human purpose and action." They allow people to experience new ways of living and feeling together, and practice habits that anchor them in new temporal trajectories.

Alternative visions of the future are particularly vital in a political and technological context where imagination is limited by the totalizing perspectives of contemporary systems of domination. Erik Swyngedouw, writing of the "postpolitical" city, describes how an artificial consensus on the definition of environmental problems "forestalls the articulation of divergent, conflicting and alternative trajectories of future (urban) environmental possibilities and assemblages" (2009, 610). Within neoliberalism, sustainability has become an "apolitical" discourse that avoids the question of "what is to be sustained and what is not?" (Greenberg 2013, 56). Contrasted with hegemonic futurity, concrete utopia is as impactful as any other prediction, plugging its audience into a new temporal trajectory that offers an alternative to the future-as-currently-planned.

Experimental infrastructures show us that "another world is possible." Avery Gordon identifies this power in the "usable" utopias described by the novelist Toni Cade Bambara: "Bambara does not create pictures of perfection, only stories of dealing with the degradations and contradictions of exploitation, racism, authoritarianism . . . differently; stories of living better than all that" (Gordon 2004, 196). Composting in community is a mode of "dealing . . . differently" with the "degradations and contradictions" of the city's food and waste systems, and other structures that produce oppression and inequality. Waste infrastructure litters the landscape of Brooklyn—mountains of trash bags, noxious transfer facilities, collection trucks. My commute by bike is filled with unpleasant smells and diesel exhaust. In Know Waste Lands, though—BK ROT's base of operations—waste infrastructure is reframed and reimagined. Waste processing shares space with a native wildflower garden. Workers play with the garbage. People walk in off the street to see how their own personal projects might be connected to the work of BK ROT. In every conceivable way, BK ROT "deals differently" with waste and space. Designers, users, maintainers, and visitors experience new sensations and feelings that can root us in something different from our daily life. Community compost sites immerse you in the smells of rotting food and fresh earth, the bodily motions of chopping scraps and shoveling compost, the connections with strangers over shared work, disgust, and hope.

The feelings we get from inhabiting the world differently—particularly hope—help us challenge taken-for-granted narratives about the future. With the affective turn in contemporary theory, a number of writers have examined feelings as active elements of social structures (Cvetkovich 2012; Berlant 2011; Kathleen Stewart 2007; Ahmed 2004). Many scholars have also examined the role of feelings in social change: Audre Lorde describes feelings as "sanctuaries and fortresses and spawning grounds for the most radical and daring of ideas" (2007, 37). Our emotional experiences of experimental infrastructures can protect ideas from the sometimes-overwhelming logic of dominant systems, and create space for reinterpreting our worlds. Hope and desire for a new kind of world makes it possible to think differently about urban environments.

Even the unpleasant affects are fecund humus in which to nurture new ways of feeling and being in the world. Because I'm studying waste, I've taken time to notice disgust, which is a rich part of the landscape of community composting. I have tromped through hot muck and pried twist ties off of semi-liquefied cilantro. The worst smells of a compost pile are thick and sticky; they linger on my clothes and in my facial hair long after the volunteer shift is over. Community composting doesn't flip the script and say that gross is good. The need for community buy-in, along with an awareness of the history of environmental injustice, requires strict discipline around the aesthetics of compost piles. Composters must be willing to sit with feelings of disgust, read maggots as messages, and manage the process of decomposition. Being present with disgust allows people to explore new forms of living with waste.

Reading community compost as utopian brings to the surface affective and experiential dimensions that are left out of traditional accounts of sociotechnical transitions. When you leave an experimental infrastructure (if you leave!), you bring along the visions and feelings you were able to cultivate there. Experimental infrastructures allow people to design and build a different future. Casper Bruun Jensen and Atsuro Morita describe infrastructures as "ontological experiments," where "multiple agents meet, engage, and produce new worlds" (2015, 85).

None of this negates the problem of busy-ness that MacBride points out in community-based zero waste reuse enterprises. While these systems "foster clean, safe, green urban communities and jobs," they are "deeply insufficient to address the problems of solid waste, its disparate effects on people of color and white working people, and the stated goals of recycling in terms of dealing with resource depletion and climate change" (2011, 128). "Small steps" toward sustainability don't necessarily add up to the kind of change that is necessary in order to build a sustainable world for everyone. Combined, the hundreds of community composting infrastructures in New York City only divert a tiny fraction of the city's organic waste. And while this might be a "small step" toward the creation of local composting systems that can manage a more significant portion of the waste stream, there are a wide variety of logistical, legal, financial, and institutional roadblocks that will make the rest of the journey difficult.

What utopia can do, though, is offer a vision of a possible future alongside the sensory, social, and emotional experiences that are necessary to knock us out of hegemonic futurity and into new realms of possibility.

The technology at the center of this book—the compost heap—has helped me think about what the piecemeal utopias of community composting can contribute to the large-scale transitions that are necessary to create just and sustainable cities. Donna Haraway develops composting as a metaphor for the ongoing process of collaborative transformation that is necessary to "staying with the trouble" of the Cthulucene:[7] "The unfinished Cthulhucene must collect up the trash of the Anthropocene, the extermination of the Capitalocene, and chipping and shredding and layering like a mad gardener, make a much hotter compost pile for still possible pasts, presents, and futures" (2016, 57). For Haraway, composting is a practice of combining existing concepts and beings and relationships to create a fertile material-semiotic humus that can nourish new life.[8]

Composting is an approach to utopia that grounds our stories about possible futures in the storied worlds we already inhabit. Compost utopianism acknowledges and reinterprets the past in order to imagine futures that make sense to us in the present. This is in contrast to a style of utopianism that does not compost, where the past and present can be forgotten in favor of prefabricated futures. Noncomposting utopianism—let's call it landfill utopianism—is

7 Cthulhucene is Haraway's alternative to the Anthropocene, named after a fictional octopus-headed cosmic entity.

8 Jennifer Mae Hamilton and Astrida Neimanis develop Haraway's metaphor as an ethos for feminist scholarship in the environmental humanities: "Composting explicitly values and deliberately repurposes extant matters—namely feminism—in the search for different kinds of worlds" (2018, 505).

an approach to the future where the past is disposable and a better world can be built on top of it using raw materials. This can leave the toxic legacies of the past and present unaddressed. It also throws away resources that are currently available, in favor of a future that might be meaningless to the people of the present. A landfill is unstable ground to build on.

Compost and landfill utopianism aren't mutually exclusive; most utopian projects have elements of reuse and elements of disposal. The New York City municipal organics pilots are landfill utopianism insofar as DSNY imported a preformed idea of what municipal compost could look like, and tried to implement it in a context where only a fraction of the city had any interest in the project. But DSNY also composted the ideas and structures of community-driven projects, incorporating them in service of creating a "culture of composting." Community composters likewise employ both styles of utopianism: while they compost the visions and values of their communities into new ways of organizing their worlds, it's hard to build systems that include everyone's ideas.

METHODS

CASE STUDIES

When it was announced in 2013, the New York City curbside composting program was projected to be the largest municipal organics collection system in the world (DSNY 2016). I was in my second year of a PhD program, and had been building a dissertation project around food waste. I was curious about how the compost pilots would interact with other approaches to food waste that were already active in New York City: large-scale food rescue organizations, dumpster divers, and community composters. I moved to New York City from the Hudson Valley to see what I could learn on the ground about how these systems were interacting with municipal curbside composting. After several months of food rescues and dumpster runs in New York City, I decided that the interactions between food rescue and composting were not as interesting to me as were the interactions between different infrastructures for composting.

I started to catalog the different organizations that were building compost infrastructures in New York City. I excluded large-scale commercial composting because of the notorious secrecy of the waste industry (Colombijn 2020; Feldman 2018). I also chose to exclude backyard composting because it seemed to be a fairly minor part of the system, in terms of both material impact and social visibility. This left me with the municipal curbside organics pilots, community composting, and compost microhaulers, which accounted for the majority of organics recycling in New York City's residential waste stream.

Studying infrastructures involves foregrounding typically backgrounded work, in a move that Geoffrey Bowker and Susan Leigh Star call "infrastructural inversion" (2000). This analytical move is not necessarily straightforward. Infrastructures are infrastructured by other infrastructures, so the process of defining foreground and background involves choosing where to cut. The compost infrastructures in New York City define the backgrounds against which other compost infrastructures operate, so I want to account for the analytical decisions I made in my effort to isolate case studies from the overlapping infrastructures of waste management in New York City.

Star suggests that an ethnography of infrastructure should identify *master narratives* in infrastructures and how they exclude others (1999). I defined infrastructures around relatively cohesive master narratives about the ways that users and workers interact with waste systems in their daily lives, and about the future of urban composting. The case studies meander through occasionally overlapping organizations and systems, following the master narratives that bring them together into infrastructures.

The NYC Curbside Organics Pilot was a municipal effort to build an infrastructure for citywide organics recycling in New York City. The pilot had been mandated by Local Law 77 of 2013, and coordinated through DSNY's Bureau of Recycling and Sustainability. The long-distance hauling and processing infrastructure was mostly built through contracts with commercial haulers—Waste Management, WeCare Organics, and Brooklyn Transfer—and processing facilities in New York, New Jersey, Connecticut, Massachusetts, and Delaware. DSNY also provisionally partnered with the New York City Department of Environmental Protection and the New York City Department of Parks and Recreation (Parks Department) to pursue possible strategies for collecting and processing organics and distributing finished compost. To describe how the Curbside Organics Pilot brought together all of these elements, I traced flows of organic waste and compost visions around New York City and beyond.

It was less straightforward to figure out how to approach community compost, which comes in many shapes and sizes in New York City. The

DSNY Community Composting Report identifies sites that process anywhere from 10 to 70,000 pounds of organic waste per month (2014). As I tried to identify representative examples, I was surprised to find a large degree of similarity among community composters, owing largely to the influence of the NYC Compost Project. The Compost Project is an office within DSNY that funds staff at eight midscale community compost projects, runs the Master Composter Certification Course, operates compost drop-off at greenmarkets, coordinates collaboration across community compost sites, provides material support to community composters, and hosts an online map of compost drop-off sites in New York City. I decided to center my analysis of community compost on the Compost Project because it plays a large role in shaping community compost systems. I investigated the project by working with and interviewing staff at four Compost Project–funded sites: Big Reuse, the Lower East Side Ecology Center, Earth Matter, and Red Hook Farms; and two unfunded sites: Compost for Brooklyn and another Brooklyn garden that has asked to remain anonymous. I also attended Compost Project events, spoke with program staff, and volunteered with food scraps collection at farmers markets. I applied for the Master Composter Training Program twice, and was rejected twice.[9]

When I started this work, there were three compost microhaulers in New York City: Vokashi, Common Ground Compost, and BK ROT. While these organizations shared a sense of solidarity, they varied widely in terms of their values, practices, and organizational structures. I became particularly interested in BK ROT, a small nonprofit organization that conducts bike-based, youth-powered food scraps collection from households and businesses in the largely Black and Latinx but rapidly gentrifying neighborhood of Bushwick, Brooklyn. BK ROT charges for collection and pays wages that start at 15 dollars an hour.[10] The founders of BK ROT, Sandy Nurse and Renée Peperone, have centered an environmental justice (EJ) approach to compost, and strive to build a form of waste management that creates sustainable local food systems and good, safe, ecologically sound jobs for youth of color. Because of its small size, it was relatively easy to draw a boundary around BK ROT, based on the day-to-day activities of its organizers and

9 I assume that I was rejected because I was not embedded in any specific community in NYC; it was not clear I'd have the kind of social capital necessary to make a community compost site meaningful to my neighbors.

10 This was before the successful "Fight for $15" campaign to increase minimum wage in NYC to 15 dollars an hour. Wages now start at $15 per hour, and are increased to $20 per hour after three months.

workers. I have also tried to account for its relationships with city agencies (DSNY, Parks Department, the Business Integrity Commission) and activist organizations (Mayday Space, The Laundromat Project).

DATA COLLECTION

Dorothy Smith articulates guidelines for *institutional ethnography*, a feminist methodology for examining how social relations work through women's everyday lives (2005). By beginning our inquiries with daily life, institutional ethnography allows researchers to understand the ways that institutions and discourses coordinate actions across a variety of different lifeworlds. Paying attention to the way that frameworks such as "ecological citizenship," "zero waste," or "environmental justice," or texts such as the NYC Master Composter Manual, organize activities across my field sites helps me to relate the structuring of the waste stream to the structuring of daily life. A sensitivity to the daily routines of waste labor and household disposal helps me to identify the institutions that structure this work.

The bulk of the data collection included participant observation at volunteer sessions. Volunteering is a central part of the grassroots waste systems of New York City; all of these systems rely, to some degree, on volunteer labor. Volunteering offered a useful entry point into research fields: with minimal training, it allowed me to see and feel the everyday workings of organizations, to understand the forms of labor they depend on, and to speak to other volunteers and to understand how their work is meaningful to them. People sometimes feel freer talking about their hopes, dreams, and values when face-to-face with rotting food.

I conducted 26 semistructured interviews with organizers, volunteers, and project managers throughout the compost systems of New York City. In these interviews, I asked people to explain the daily work of composting and what it meant to them, and to talk about the design decisions that have gone into these systems. Interviews were recorded and transcribed, and keyword-coded for content. Six interviewees declined to be recorded; their interviews were notated using my own shorthand, and used as guides for inquiry.

I also drew from primary documents such as websites, press releases, fliers, publications, reports, promotional videos, and other materials that are publicly available— and participant observation at planning meetings, training sessions, and public forums.

DSNY posed a significant methodological hurdle in that I was largely unable to find anyone working in the Bureau of Recycling and Sustainability who was willing to be interviewed on the record, or gain access to any facilities in New York City. As such, my account of the municipal curbside organics collection pilots is based on what I had access to: reports, public hearings, Solid Waste Advisory Board meetings, news stories, visits to compost facilities, and observing the rollout of the pilots from my bicycle.

At the other end of the spectrum, BK ROT offered a remarkable degree of access, and I rapidly found myself becoming friends with the coordinators and "an authentic part of the team" (BK ROT 2016). This has allowed for a significant degree of insight on the details of their work, and I've been able to talk through much of my analysis with the organizers. However, this has also raised both ethical and epistemological issues, which I address below.

Analysis consisted of keyword-coding interview transcripts, fieldnotes, and primary documents, and identifying new themes in an iterative fashion. While I looked for themes that were shared across field sites, I decided to organize my analysis around individual field sites.

RESEARCH FEELINGS

I found myself caught up in the emotional welter of fieldwork and infrastructure building. As an academic, I felt a pressure to problematize the systems I was studying. In my graduate classes, I had voraciously consumed critiques of recycling, volunteerism, and urban gardens, and I knew what to look out for in the field. As an activist, though, I found myself mirroring the excitement of my interlocutors. At both large and small scales, composting is an awe-inspiring process. And my experience in punk collective houses in Washington, DC, had primed me for getting involved in the maintenance of makeshift systems. I could not shake either of these feelings—the thrill of critique or the fulfilling work of systems building.

Eve Kosofsky Sedgwick helped me to make sense of these conflicting feelings with her definitions of *paranoid* and *reparative readings* (Sedgwick 2003). She explains that scholars make a habit of trying to unmask the nefarious workings of power structures, often (she hypothesizes) for the pure joy of being right about their own suspicions. She describes this form of scholarship as *paranoid* reading and prescribes, as an antidote, a habit of *reparative* reading: reading that is pleasure-seeking, loving, and capable of imagining that the future might be different from the present. Reparative readers avoid the paranoid tendency of trapping themselves descriptively within an internally cohesive system of domination, and instead try to write in a way that is surprising and generative. Reparative readings are essential to studying utopia, where the assumption that another world is possible is foundational to the field. Rather than try to ditch my affective commitments in favor of an objective detachment, I've tried to use paranoid and reparative readings as emotional 3-D glasses. I correct paranoia with repair and vice versa, suggesting how these systems help to reproduce structures of power *and* how they can be used as a means of escaping them.

STANDING WITH SUBJECTS

Kim TallBear describes a feminist ethics of "standing with" the communities you study, which has helped me to understand my own positionality in the field. *Standing with* subjects challenges the ethnographic separation between researcher and subject, conceptualizing research as "an opportunity for conversation and sharing of knowledge, not simply data gathering" (2014). Following TallBear's ethical/epistemic example, I've tried to understand my research as part of an ongoing conversation in the field of community composting, rather than as a disinterested or disconnected description. This has shaped the selection of case studies, the questions I've asked, the ways I've shared findings, and my ongoing commitments. I have purposely chosen case studies that I can care for and not simply critique.

One of my motivating questions is "What do grassroots infrastructures do?" This question responds to a challenge posed by critics of small-scale solutions, who sometimes describe community composting as a minor phenomenon and not worth the resources needed to support it. Answering this question can be useful both for helping to guide the work of grassroots systems designers and for legitimating grassroots systems to funders who favor large-scale change. The concept I articulate in response, "experimental infrastructures," is meant to highlight the bottom-up knowledge production taking place in grassroots infrastructure projects.

This started as an academic project, but I've done more work to share my findings with fellow composters than with fellow academics. I have presented at industry conferences on composting and food waste, often both as an academic and an organizer. I've spent more time writing grants for BK ROT than I have writing for academic publications.

Finally, I've continued working with BK ROT long after my fieldwork officially ended. I joined its advisory board in 2016, its board of directors in

2017, and in 2019 I took on the role of board chair, which I held until the end of my term in 2021. Sandy explained that they wanted me in the role because I had spent five years recording their actions and values, and checking in with them about my understanding of their work. I have been proud to play these roles, but I also recognize that they have shaped my understanding of the organization and the ways I've been able to describe them.

Instead of trying to position myself as separate from BK ROT, I've taken on its core values as part of my account of grassroots composting. While this sacrifices analytical distance—a traditional value in social science—I believe this has also made it more possible to convey the situated infrastructural imagination that animates BK ROT. The kind of hope I'm trying to describe here is not going to come from an artificially removed vantage point with an objective perspective on the city's past, present, and future. Hope is a feeling, and standing with subjects allows me to convey this feeling more authentically.

Of course, standing with subjects does not erase the power dynamics of ethnography. As a white academic from a middle-class background, I've been able to enter the field, turn some compost piles, and leave the field (for the most part) to pursue a (contingent) academic job with a middle-class salary and benefits. Through my efforts to center the expertise of youth of color, I've been able to center *my own* expertise and earn a PhD, without offering similar benefits to the youth I worked with. I entered the field sensitized to some of the problems of representing research subjects but not to the divergent trajectories of grassroots activists and academics. I did not recognize the kinds of diligence it demands to be in right relationship with subjects. I can't offer any useful resolution of these inequities, and the confessional genre seems unlikely to address them meaningfully. But I am committed to naming this dynamic.

Other than standing with subjects, all interviewed subjects had explained to them the purposes of the project, and had the option to remain anonymous if preferred. Several informants' names have been changed in order to protect their identities. In some cases, the names of organizations have been changed.

STRUCTURE

Chapter 1 is a brief genealogy of "organics recycling" as a concept that shapes the activities of industrial compost in the United States. I describe the current position of compost as a boundary object between organic food, zero waste, and industrial waste management. This chapter argues that the technological frames that have guided the development of compost systems have tended to privilege waste management over meaningful work or holistic agricultural practices, and that the ideologies and infrastructures that dominate contemporary compost systems could have developed otherwise.

Chapters 2, 3, and 4 describe the case studies. Each case study chapter is composed of a set of interconnected narratives that articulate the social, cultural, and technical interventions taking place in the New York City waste stream. In the conclusion to each, I describe the interaction between formal and informal infrastructures, then articulate both a paranoid reading and a reparative reading of each project, working to square the two in order to provide a fuller picture of a changing system.

In chapter 2 I focus on the organics recycling pilots conducted by DSNY. Starting with curbside collection pilots in the 1990s, I describe how the city has made decisions about whether to collect organic waste, who to collect it from, how to collect it, what to do with it, and how to convince people that organics recycling is important and good. I also discuss the impacts of the COVID-19 pandemic on curbside composting. This chapter introduces many of the players in New York City's formal organics recycling system, and describes how this system is shaped by the logic of neoliberal waste management.

Chapter 3 describes the NYC Compost Project, an office of DSNY that has a hand in community composting citywide. I share histories of two Compost Project "host sites," midscale DSNY-funded compost operations with a community focus. The rest of the chapter is structured around themes that came out of interviews and participant observation, including questions about labor, expertise, data collection, and the "community" of community composters. I conclude by describing how the utopias of community composting serve the needs of both DSNY and community activists.

Chapter 4 describes the work taking place at BK ROT, a compost microhauler built around the principles of just sustainability. I articulate the specific ethos of BK ROT's approach to composting, and explain how it has pursued its social justice values in several different areas of work: addressing gentrification, odor management, community building, visioning, building infrastructure, and coordinating labor. I then return to the metaphor of "composting utopia" in order to describe how BK ROT responds to its context.

In the conclusion, I pull from each of the case study chapters to explore three main takeaways from this book. First, I describe how neoliberal waste management has shaped both municipal and community composting in New York City, and how people have innovated within neoliberal waste management to develop alternate approaches. Then I examine how experimental infrastructures play a role in the city's transition to composting, by enabling new forms of response-ability, assembling new communities, and building partnerships with municipal institutions. I conclude by discussing the specific relationship with hope nurtured by composting, and explain what this can offer to the pursuit of utopia.

In the face of interlocking social and ecological crises, the need to build sustainable systems is urgent. However, the dominant strategies for working toward sustainability have been limited by the depoliticized rationality of neoliberalism. In experimental infrastructures, organizers enact practical utopias that have the potential to articulate alternative, just sustainabilities and cultivate hope for better futures. It's my hope that the stories in this book will offer models for you to compost your own utopias, finding new possible futures in the fragmented present.

CHAPTER ONE

A Genealogy of Municipal Organics Recycling

With the curbside organics collection pilots, New York City joined a growing list of municipalities in the United States that ask residents to separate banana peels and used coffee grounds from the rest of their trash. Between 2005 and 2017, the number of municipal food scraps collection programs increased thirteenfold, from under 25 to 326. By 2017, 5.1 million households in the United States were served by source separated organics collection (Streeter and Platt 2017).[1] Many of these programs operate in cities that have a history of green policy and infrastructure, such as San Francisco, Seattle, and Portland (Schouten, Martin, and Tillotson 2015). A handful are in smaller cities where disposal is difficult or expensive: for instance, Mackinac Island in Michigan, a car-free tourist destination, has collected food and yard waste by horse-drawn buggy and composted it locally since 1992 (Yepsen 2015). Efforts to implement composting in New York City feel particularly meaningful, given both the complexity of the effort, and the city's reputation as a trendsetter. DSNY proclaimed in a series of advertisements in 2019, "If you can compost here, you can compost anywhere."

The 1.8 million tons (N. Goldstein 2017a) of food waste that the United

1 Exactly what this stream can consist of is somewhat flexible on a city-to-city basis, based on the kind of composting system being used; collection habits, the city's concern with rats and flies, or with the safety of the finished product; and the city's own ongoing negotiations with compliance. Some systems collect only raw fruits and vegetables, while others include cooked foods, meat and dairy, and soiled paper products. Many include yard waste as well. I have not found any evidence of US cities collecting diapers or sanitary pads, but Toronto has collected these products since 2009 (City of Toronto 2017; Galbraith 2009).

States composts annually make up a sliver—0.3%—of the 665-million-ton estimate for annual food waste in the United States (Buzby et al. 2009). Residential food scraps collection is not yet standard in the United States, but advocates have described it as "the next frontier" of recycling (G. Smith 2020; Stein 2017; Ferro 2016; Carpenter 2014). Municipal organics recycling promises to reduce the amount of waste sent to landfills, create compost for agriculture and landscaping, create renewable natural gas, create jobs, and potentially reduce the carbon outputs of our waste systems (Bradford et al. 2019; Platt et al. 2014; Pinto and Barnes 2013; Platt, Bell, and Harsh 2013). When I separate out my food scraps, I am performing my affective investment in a zero-waste future.

Recycling systems are often designed to inspire hope (MacBride 2011; Hawkins 2006). City officials wrap organics recycling into feel-good progress narratives in which cities are reducing their ecological footprint by cutting down on waste (City of New York 2015), creating clean energy and returning nutrients to the earth (City of Cambridge 2019), and protecting local ecosystems from urban pollution (Seattle 2019). For the proponents of municipal compost, organics recycling—like paper, plastic, metal, and glass recycling before it—is the result of a wasteful society wising up about its flagrant misuse of resources, and turning proper wasting into a productive, civically minded activity (MacBride 2011).

While recycling has the potential to reduce human impacts on the environment, recycling systems often represent a compromise between environmental principles and industrial interests, and many are insufficient to manage the wastes they respond to (MacBride 2011; Lounsbury, Ventresca and Hirsch 2003; Pellow 2003). Samantha MacBride argues that the current practices of recycling are not in place because they are the most effective ways to reduce waste, but because of corporate and state interest and the perceived inflexibility of waste generators (2011). Rather than alleviate the social and environmental burdens of waste, recycling in the United States has entrenched an ineffective form of waste reduction that tends to place the burden on cities and residents (in the form of curbside collection and education) rather than on producers (in the form of bottle bills, container taxes, or extended producer responsibility). Citing the low national average recycling rate (33%) alongside increased waste generation, MacBride is dubious of the claim that "progress is being made" toward waste reduction: "In a sense, progress is indeed being made; it is just that this progress is one of state program building and associated economic activity, not material progress in reducing waste or social progress in making the burdens and benefits of waste disposal/diversion more equitable" (83). Organics recycling is progress, in a sense. But progress toward what?

This chapter traces a genealogy of municipal organics recycling in order to situate New York City's curbside organics pilot in a historical trajectory that shapes policymakers' and activists' compost imaginations. There is reason to be hopeful about municipal compost, but the dominant approach available in the United States is historically specific, and offers a limited vision of the ways that food waste can be managed in cities. Organics recycling often refers to a specific, centralized form of disposal in which "waste generators" separate out organics at curbside for collection by sanitation trucks, which move the material to a handful of industrial facilities that systematically decompose large volumes of material, mostly for landscaping uses, if for any specific use at all. The dominant approach values volume and diversion over other possible functions of composting, such as closing local resource loops, replenishing soils, or creating good jobs. As a solution to food waste, it treats uneaten food as a given rather than as a symptom of structural problems in the food system.

None of these values are inherent in organics recycling or composting. But large-scale centralized collection and processing of source-separated organics has become the standard, depoliticized vision for the "next frontier" of recycling within neoliberal waste management. Many features have been determined by ongoing efforts to establish markets for collection and processing of organic waste and distribution of finished compost, rather than by environmental measures alone.

Genealogy defamiliarizes the taken-for-granted elements of contemporary discourse (Kendall and Wickham 1999). A side-effect of this is to recognize alternative trajectories that could have been taken. José Esteban Muñoz describes a Bloch-inspired practice of "putting the past into play with the present" (2009, 28), in which the past can be a source of hope that challenges the "tautological nature of the present" (28). Tracing this specific genealogy of organics recycling reveals a variety of alternative values and visions that might be useful for reimagining our waste systems. Some of these values are already in play at both small and large scales as community composters, microhaulers, and even some municipalities keep resources local, create closed loop systems, and build for community empowerment.

MUNICIPAL WASTE MANAGEMENT MAKES GARBAGE DISAPPEAR

Households have always generated waste, but the idea that waste should be managed by cities is relatively recent in US history (Melosi 2005; Hickman 2003). Before the US industrial revolution, a largely rural population with plenty of stolen land for dumping made it possible for many US residents to generate waste without worrying about where it would end up (Melosi 2005).[2] Even in cities, households had uses for many of the materials that would later make up the waste stream. Susan Strasser documents the resource recovery work of US women in the 19th century, who used organic remnants as fuel, fed it to domestic pets or livestock, or rendered it for other purposes (1999). Peddlers established loose barter networks through which women could trade rags, bones, and many of the items that they could not reuse. While residential waste existed, and piled up on city streets, it was not understood as a social problem or a responsibility of cities.

The geography and economics of household wastes changed during the US industrial revolution. The concentration of factories drove tens of millions of people into the cities over the 19th century, increasing the urban population nearly thirtyfold. More people meant more waste, and a problem that had once been a nuisance was reframed as a public health hazard (Melosi 2005). Refuse and garbage in the streets became the objects of informal economies: scavengers collected rags, bones, and manure off the streets to be made into paper, gelatin, and fertilizer (Strasser 1999).[3] But informal scavengers were unable to manage the massive wastes produced by the growing cities. Rags, bones, and ash, as well as more abject stuff—human and animal excrement and abandoned dead horses—piled up and rotted in the streets (Nagle 2013; Melosi 2005). Urbanization also moved more waste generators away from farmland where their organic waste could be useful for feeding animals and soil (Blum 1992).

In the late 19th century, sanitary reformers organized to clean up the waste problems in the cities. There were plenty of aesthetic reasons to organize around waste, on top of the necessity of passable streets. But hygiene was the primary battle cry of the early advocates of US waste management. Refuse was understood to be the source of disease, and keeping the streets empty

2 Claiming land for waste has been an important element of colonization. Max Liboiron's excellent *Pollution Is Colonialism* (2021) explores this theme with depth and insight.

3 These underground economies should not be romanticized: Catherine McNeur documents the use of brewery waste as feed for cattle in NYC in the mid-19th century, which resulted in a nutritionally impoverished "swill milk" that left many children undernourished (2014).

was a matter of public health. Cities responded to the waste problem by contracting with private scavenging companies and developing their own street cleaning departments. By 1880, the majority of large US cities had some form of sanitary authority who had oversight over street cleaning (Melosi 2005).

It was not immediately clear what should be done with refuse once collected. Disposal was the primary goal of street cleaning, and many US cities resorted to simply dumping their wastes on land or in water (Melosi 2005). Some cities explored incineration as a means of reducing the volume of wastes and making them more manageable (Strasser 1999; Hickman 2003). For sanitary reformers, fire was appealing as a "disinfectant" for unhygienic wastes, effectively neutralizing the threat that garbage posed to public health (Melosi 2005).

Detractors as early as the 1890s claimed that incinerators and dumps were a waste of potential resources (Strasser 1999; Melosi 2005). Many cities pursued recycling and reuse efforts in order to recoup some of the costs of street cleaning (Melosi 2005). Cities would also create markets for managing urban detritus. New York City coordinated a system for manure disposal in which independent scavengers collected manure from the streets to be composted with sawdust and tannery by-products and sold to farmers outside the city (Goldman 1997). Colonel George E. Waring, New York City's commissioner of street cleaning from 1895 to 1898 and the architect of the city's early sanitation system, implemented a source separation program that asked residents to separate organic waste, rubbish, and ashes into different containers so that street cleaners could handle wastes more easily. Source separation was not immediately popular among New Yorkers, but it set a precedent for a shared responsibility between city and residents for the management of wastes: residents sorted their wastes and kept them neat, and the city collected them (Melosi 2005).

Reclamation met challenges. The rise of consumer products created a more complex waste stream, which was more difficult to manage at the household and city levels, and less profitable to mine for organic resources (Strasser 1999; Blum 1992). And there were technical problems with many of the reclamation methods. Pig farms were a popular destination for organic waste, but were also a breeding ground for *Trichinella spiralis*, which caused outbreaks of trichinosis (Hickman 2003). Reduction plants heated organic wastes to create a residue that could be used as fertilizer, but these systems were expensive to run and needed to be placed far from populated areas due to the odors they produced (Melosi 2005; Strasser 1999).

More damning was the fact that waste reclamation was not as lucrative as its proponents had hoped, and failed to compete financially with cheaper disposal technologies. In 1937, the first sanitary landfill was built in Fresno, California. Landfilling offered a cheap and effective solution to wastes, and rapidly grew in popularity after World War II (Hickman 2003). And improvements to garbage incineration in the 1930s and 1940s allowed many cities to pursue incineration as a way to reduce wastes before landfilling.

Landfilling and incineration were able to address the problem of urban waste disposal, and helped to cement the hygiene movement's framing of the waste problem: *the garbage must go*. Urban waste management came to be directed at efficient disposal rather than materials management. This historical beat also established roles for urban waste management and residential waste generators: it became the city's job to coordinate waste disposal, and the residents' job to set out their waste for tidy collection.

COMPOST CAN REPLENISH DEPLETED SOILS

There is archaeological evidence that cultures all over the world developed agricultural practices that involved returning organic material to the soil through guided decomposition. In Neolithic Scotland, farmers composted household wastes in middens, which they plowed to plant their crops (Guttman 2005). In the Amazon basin, starting around 450 BCE, Indigenous farmers created the fertile, long-lasting soil known as *terra preta* (black earth) by burying charcoal, farm wastes, and manures (Woods and McCann 1999). In China, farmers tilled weeds and human and animal manures into the soil as early as the 3rd century BCE (Wen 1984). In Ghana and Liberia in the 13th century CE, women returned ash, bones, soap-making byproducts, farm residues, and food scraps to the ground to create the carbon rich soils known as *African dark earth* (Penniman 2018).

In order for Europeans and Americans to re-“discover” composting in the 20th century, it needed to be forgotten. The urbanization of Europe in the 18th century and the United States in the 19th century disrupted older traditions of closed-loop farming. The same processes that concentrated wastes in the cities also removed them from the countryside, where human, animal, and vegetable waste had previously been used to replenish soil fertility (J. B. Foster 1999). Systems for returning human waste to the countryside existed,[4] but the technologies for reclaiming wastes couldn't manage the increasingly concentrated wastes of cities (Melosi 2005). At the same time, a growing demand for food pushed at the carrying limits of the soil (Foster and Magdoff 1998). Farmers and agricultural scientists sought out new methods to increase their yields to feed the growing cities.

The development of agricultural chemistry redefined the problem of soil fertility as primarily a chemical problem: how could farmers replenish the nitrogen, phosphorus, and potassium that were responsible for soil fertility (van der Ploeg, Böhm, and Kirkham 1999)? Imported guano—dried seabird droppings—became a popular response to this question. By the time it was adopted by European colonizers, guano had been used by the Quechua-speaking farmers of what is now Peru for centuries. Starting in the 1850s, European traders took control of the country's guano resources, entrapped Chinese laborers in debt peonage in order to mine guano, and sold it to European and US farmers in a lucrative market (Melillo 2012). In Europe and the United States, imported guano—and later nitrate—was seen by many as the solution to losses in soil fertility. Many farmers abandoned the closed-loop systems that could no longer keep up with the demand for food, in favor of "open" systems that relied on imported fertilizers.

Imported fertilizers were effective for growing crops, and had the allure of technological progress, but the guano trade didn't put an end to research on and use of local organic wastes for farming. The relative merits of imported guano and homemade manures remained uncertain even as guano became the preferred fertilizer—experts recommended that farmers carefully weigh the relative costs and benefits of guano against traditional manures (Robinson 1865; "Guano as a Manure" 1852). Already, some experts recognized compost as a more holistic approach to soil health. For example, Solon Robinson's *Facts for Farmers* recommends compost both as fertilizer and pest control (1865). But in the early 20th century, imported and artificial fertilizers came to dominate the agricultural systems of the United States and Europe (Obach 2016).

The guano trade is one example of how dominant European and US relationships with the environment have relied on colonial power and white supremacy: the most widely used farming practices have been built

4 For example, in preindustrial England, "nightmen" would collect human excrement from homes and cart it to nearby farmland.

on exploitation, enslavement, and primitive accumulation of Native land and resources. Proponents of abolition ecology (Heynen and Ybarra 2021; Heynen 2020) or radical ecologies (Pulido and De Lara 2018) have argued that real environmental justice relies on directly addressing the racism and colonialism at the core of both environmental destruction and contemporary environmentalisms.

George Washington Carver's experimental station in Tuskegee was an early site of resistance against the agrichemical regime and its relationship with racial capitalism (Hersey 2006). Recognizing that imported fertilizers were beyond the financial reach of Black sharecroppers, and that reliance on guano forced farmers deeper into debt, Carver advocated the use of readily available organic materials—particularly swamp muck—that could be decomposed to produce fertilizer. Instructions he provides in his 1917 pamphlet, "Twelve Ways to Meet the New Economic Conditions Here in the South," are similar to contemporary composting techniques. He describes bin and windrow systems with alternating layers of carbon-rich and nitrogen-rich materials, and flexibility about what could be included: "wood ashes, old plaster, waste lime, rags, paper, or any matter that will decay quickly" (4). While Carver's writings helped many Black sharecroppers escape the cycle of debt and own their farms, his ideas about compost did not engender a large-scale transformation in US agriculture. Mark Hersey explains this as the result of white landowners' resistance to Carver's liberatory politics of agriculture, and the increasing power of the industrial, chemical-based agribusiness system (2006).

In the 1930s, Albert and Gabrielle Howard brought a more abstract set of values to the agricultural use of compost by framing it as part of a morally superior form of agriculture, more in line with natural resource cycles. From their experiences studying agricultural techniques in India, the Howards learned that the productivity and disease resistance of crops depends on the health of the soil, and that returning precisely decomposed plant matter to the soil is necessary to maintaining its health. They brought these techniques back to England as "the Indore process," named after the region of India where they learned it from farmers. The soil science that Howard and his collaborators developed challenged the chemical model of plant nutrition as reductionist, and suggested new directions for understanding soil as a living thing (Gieryn 1999).

The myth making around the Howards has shaped the ideology of compost. Albert Howard is widely considered "the father of modern composting," and histories of compost often jump straight from ancient farming techniques into the "invention" of the Indore process, and the Howards' efforts to popularize composting in Europe (Golueke 2009; Blum 1992; Hepperly and Ziegler Ulsh 2007; Earth Matter 2014). But given that US farmers recognized compost as an alternative to fertilizers long before the Howards went to India, it seems that the Howards were not technical but cultural innovators. Earlier agricultural scientists had described the ability of compost to control pests and provide lasting soil fertility. Even the Howards' cosmology of compost was not new; decades earlier, Carver had "emphasized the "organic unity" of the universe" (Hersey 2006, 245) and had claimed that "nature does not expend its forces upon waste material, but that each created thing is an indispensable factor in the great whole" (quoted in Hersey 2011, 44). Only a handful of histories of compost emphasize Carver's contributions (Tilth Alliance 2019; Penniman 2018; Martin, Gershuny, and Minnich 1992).

Why has Howard become the preferred origin story for composters, while Carver has fallen by the wayside? Both scientists had revolutionary visions of agricultural reforms, and both had only minor impacts on agricultural practices within their lifetimes. Both envisioned composting as part of an agriculture that was more in tune with the natural order. It's possible that the difference lies in the political foundations of the two scientists' compost advocacy: for Carver, composting was a technique for making do with what you had, in an effort to escape the economic power of landowners and the fertilizer industry (Ferrell 2010; Hersey 2006). It was part of what Nik Heynen characterizes as "abolition ecology": an ecological imagination grounded in challenging racial oppression and enabling Black and Indigenous people to shape their own relationships with labor and land (2020). Compost was an affordable alternative to expensive fertilizers that offered Black sharecroppers the promise of independence (Carver 1917, 1936).

For Howard, composting was primarily a method of building a holistic relationship with the soil (2010). The growth of the synthetic fertilizer industry following World War I meant that nitrogen-rich fertilizers were cheaply available, and Howard had to defend the use of compost in the face of cheaper alternatives that damaged the soil (Gieryn 1999). Howard's relatively apolitical vision of compost frames the practice as good for soil, and for the earth, but not necessarily for farmers.

While Howard did not create a widespread shift in farming, a small movement of organic gardeners picked up Howard's ideas and used them as the foundation for a new kind of agriculture. Howard's account of the importance of composting became the standard for the composting industry that would start to develop in the 1960s. The exclusion of Carver's liberatory politics of composting from conventional accounts means that the dominant discourse treats composting and compost application as virtuous deeds, in line with an overarching vision of earth systems, rather than practical solutions to the socioeconomic problems of agriculture, or a challenge to racial capitalism.

COMPOST CAN MANAGE MUNICIPAL WASTE

Composting was developed as a waste management technique independently of the organic movement. Even before the Howards brought the Indore method to Europe, European engineers were developing processes to compost manure and municipal waste for hygienic disposal (Hickman 2003; Golueke 2009; J. W. Smith 1924). Giovanni Beccari patented the Beccari process in 1920, which begins with a month-long anaerobic fermentation in a sealed container, followed by 15 days of curing. In 1930, Danish engineers developed the Dano Process, which involves agitating mixed materials (organic and nonorganic waste) in a rotating, heated drum for 5 to 10 days and then curing them in compost heaps (Hickman 2003). Variations on these methods became popular across Europe, where space for landfills was limited and farmers could easily incorporate compost into their fields (Blum 1992).

Municipal compost did not have such a promising beginning in the United States. In the 1930s and 1940s, US cities experimented with European composting methods, but they weren't as effective with American refuse or agricultural practices. Five Beccari plants opened in the United States during this period, but all were later abandoned (Blum 1992). Later engineers attribute the "dismal start" (E. Epstein 1996, 13) of compost in the United States to the variability of its waste stream, the apparent lack of need for organic fertilizers, the duration of the process, and distrust in locally untested technology (Blum 1992; Hickman 2003). Industrial composting was neither compatible with US agriculture nor with US garbage.

Growing US cities after World War II strained existing systems for managing urban wastes, and municipal engineers again posited composting as a partial solution to the new waste problems. This time, American engineers needed to develop composting methods that worked for American trash. Research teams at the University of California at Berkeley, Michigan State

University, and the Public Health Service in Georgia implemented pilot projects to examine the feasibility of large-scale composting of American waste and the nutrient density of the resulting product (Hickman 2003; Golueke and Diaz 1996). Clarence Golueke of the Sanitation Engineering Research Laboratory at UC Berkeley describes the rules of municipal and agricultural composting that were delineated in the 1950s and early 1960s:

> 1) The process should be aerobic, in order to reduce odors.
> 2) The process should be able to produce heat on its own.
> 3) While previous engineers had inoculated materials with bacteria to speed up the process, these inoculants are not necessary.
> 4) Environmental factors are unpredictable, so absolute standards are hard to come by.
> 5) Materials will need to be allowed to compost at least two weeks before they are no longer a nuisance. (2009)

While the broad strokes are similar, midcentury industrial composting practices were different from the previous guides to small-scale agricultural composting. Carver had specifically instructed farmers not to allow compost heaps to get hot enough to produce steam, which he believed would reduce their nutrient value (1936), and Howard had suggested that composters inoculate an inactive pile with material from an active pile (2010). Both agronomists had suggested that the process of composting should take months, not weeks. In contrast to the agricultural techniques, the principles of composting as a waste management strategy were developed in tests of mixed refuse and sewage sludge composting, and were intended to create a process that could proceed quickly, unobtrusively, and create a stabilized, hygienic product (Goluecke and Diaz 1996). Sanitary engineers did not clarify the role of compost in maintaining soil health. Marketers were uncertain if compost had any use as a soil amendment, though one writer suggested it was "an excellent vehicle for chemical fertilizers" (Cassels 1971).

Industrial composting differed most starkly from Carver and Howard's versions of composting in terms of the feedstock. From the 1930s into the 1990s, most industrial composting facilities in the United States took in mixed municipal solid waste (MSW), which includes both compostable and noncompostable materials. In a mixed MSW composting system, waste is shredded then agitated and aerated in an enclosed container so that the organic portion—including paper, cardboard, and textiles—decomposes into compost (Hickman 2003). This mix of composted organic and nonorganic waste is then screened so that the noncompostable components can be recycled or buried in a landfill. While mixed MSW composting simplifies collection, the finished product is likely to contain higher levels of heavy metals and particles of plastic and glass, which has made mixed MSW compost unpopular for agricultural or landscaping applications (Goldstein and Steuteville 1994).

The launch of the magazine *Compost Science* (later renamed *BioCycle*) by Rodale Press in 1960 brought the organic ideals of Albert Howard into conversation with the practice of composting as waste management. *Compost Science* founder Jerome (Jerry) Goldstein started the magazine while he was executive editor of *Organic Gardening and Farming*, which had advocated for "the return of humus to the soil in order to build and maintain soil fertility" (N. Goldstein 2009a, 58). In the inaugural issue of *Compost Science*, Goldstein explained that "our editorial objective is to set up a central medium—a 'clearinghouse'—for valuable information for municipal and industrial officials responsible for the treatment and disposal of organic waste materials" (quoted in Olds 2009, 44).[5] In *Compost Science*, scientists, engineers, waste managers, and organic advocates shared news stories about composting technologies and collection systems around the world. The magazine published results from a growing field of research on solid waste composting, funded by the Solid Waste Act of 1965, the USDA Soil Conservation Service, and land-grant universities. Contributors hashed out debates around the safety of biosolids,[6] the importance of mechanization, and the challenges of establishing a compost industry (N. Goldstein 2009a). Their advertisements promoted new composting equipment developed by organic farmers and waste management (N. Goldstein 2009b).

Despite the efforts of *BioCycle* and the sanitation engineers, industrial composting did not become popular as a municipal solid waste management technique in the United States at any point in the second half of the 20th century. Eighteen US cities opened plants for composting municipal refuse between 1950 and 1970, and all but two shut down before 1970 (Blum 1992).

---

5 Jerome Olds is a pen name used by Jerry Goldstein.

6 While I do not explore it at length here, the history of biosolids has had a productive interaction with the history of MSW composting. As early as the 19th century, engineers were exploring anaerobic digestion as a means of treating sewage to reduce pathogenic bacteria (Meynell 1976), and AD has been widely applied in sewage treatment in the United States since the 1980s (Marchaim 1992). As I'll discuss in the next chapter, AD has become an appealing form of organics recycling because it has the potential to produce methane, and some sewage treatment facilities have been reframed as a potential part of New York City's solid waste infrastructure. Additionally, the biosolids industry has had to do considerable work to sell biosolids as safe and useful for food crops, especially when dealing with urban biosolids (Abumrad and Krulwich 2013; Schneider 2011). These debates have paralleled evaluations of the safety and usefulness of MSW compost.

Golueke blames the failures of this period on the overmechanization of processing and overpromotion of finished compost. Promoters placed emphasis on costly equipment, when the process itself could be relatively simple. And by overselling the profitability of compost, promoters framed composting as a means of making money, rather than as a waste disposal technique. This became a problem when "the success of composting [was] measured on the basis of financial returns from the sales of the product; whereas other 'disposal' processes are judged solely by their effect on the environment" (Golueke 2009, 29). Selling finished compost worked in Europe due to financially supportive governments and the prevalence of labor-intensive agricultural practices that could make use of compost, but the same economics couldn't be made to function in the United States, where the free movement of waste across state and national borders kept landfilling cheap,[7] and compost was unpopular (Blum 1992). The majority of MSW compost produced in the United States has gone to landscaping or stormwater embankments (Gorrie 2015; Freeman, Skumatz, and D'Souza 2012).

In part, compost was difficult to market in the United States because of the specific dynamics of US agriculture. Over 99% of farms in the United States use conventional techniques, and make use of chemical fertilizers (USDA 2019). Compost application is at odds with the economic structure of farming in the United States: many farmers need immediate results because they lease their land, and it takes several years for compost to improve the soils. While the growing organic farming industry offered a potential market for compost, the fact that most compost in the United States was made from mixed MSW until the late 1990s made it both unappealing to farmers and unqualified for USDA organic certification (Relis and Levenson 1997). And even though compost was central to the early definitions of organic farming, organic certification did not require that farmers use compost. As a result, organic farms in the United States have tended to use more animal manures than MSW compost (Pires et al. 2018).

In a 1999 editorial, *BioCycle* editor Nora Goldstein laments that "composting has been forced into the realm of waste management" (4). This

7 The 1994 Supreme Court decision in *Carbone v. Clarkstown* further cemented availability of cheap landfilling. One method that states and municipalities had been using to establish markets for compost and recyclables processing was "flow control," in which private haulers were required to haul waste to specific local facilities. In *Carbone v. Clarkstown*, the court determined that flow control violated the commerce clause of the US Constitution, and prevented the formation of interstate markets for disposal. Without flow control, it was much more difficult to source feedstocks, and many compost facilities folded due to lack of inputs (Goldstein and Steuteville 1994).

observation is repeated regularly in *BioCycle*, with contributors recognizing how different the field would be if it were understood from the perspective of compost production, rather than waste disposal (Ligon, Mishra, and Votta 2000; Block and Goldstein 2000; Golueke 2009). In the United States, large-scale composting was developed as a waste management technique with only minor input from agricultural scientists and farmers. Industrial compost facilities were designed to process large volumes of waste efficiently, not to improve soil fertility.

ORGANIC WASTE SHOULD BE COLLECTED SEPARATELY

Starting in the late 1980s, widespread concern about dwindling landfill space in the United States helped to create public and legislative pressure to address the "garbage crisis" (Melosi 2008). Environmentalists pushed for reforms that would reduce the amount of waste going to landfills. Bottling industries and manufacturers of disposable products worried that waste reduction policies would undercut their business models, and pushed instead for policies that were compatible with high levels of consumption. Curbside recycling collection programs were a compromise between industry and activists: they reduced the amount of waste going to landfills, and avoided making container companies take responsibility for their wastes (MacBride 2011; Lounsbury, Ventresca, and Hirsch 2003). Curbside recycling replaced alternative models, such as the community recycling centers of the 1960s and 1970s, which had taken a holistic approach to community development, and pursued recycling alongside environmental education and jobs creation (Weinberg, Pellow, and Schnaiberg 2000). The recycling programs that were created in the 1980s and 1990s crafted a cultural framework that would go on to shape municipal composting.

In the early 1990s, disposable diapers became an important rallying point in the war on waste. By 1990, diapers made up 1.6% of the US solid waste stream, and the availability of cloth alternatives made disposable diapers seem like an easy target for waste reduction policies (Gilson and Butler 2008). In 1989 and 1990, 22 states considered taxing or banning disposable diapers (Specter 1992; *Deseret News* 1990).[8]

8 It's likely that diapers were particularly appealing because of the gendered division of child-rearing labor: women are more likely to be tasked with changing diapers and cleaning cloth diapers, so the burden of a switch to cloth diapers would be mostly managed by women (Strasser 1999; Wajcman 1991). Disposable diapers have likely benefited women more than men, in terms of both convenience and cleaning labor. The

Procter & Gamble, manufacturer of Pampers and Luvs diapers, controlled around 50% of the disposable diaper market at the time (Nolan 1990). In order to protect its market, it pushed composting as a solution to diaper waste. The company launched advertising campaigns that proclaimed that its diapers could naturally break down into nutrient-rich compost—even though the diapers' plastic outer cover was not biodegradable (Holusha 1991). In 1990, P&G brought together waste industry, manufacturers, environmental activists, and regulators to create the Solid Waste Composting Council, which would advocate for the creation of municipal composting systems. At the time, mixed waste composting was still the norm for MSW composting. This was appealing to P&G because its diapers could be composted only in mixed waste systems. Proctor & Gamble put $20 million toward research and construction of mixed waste composting infrastructure (Nolan 1990).

Proctor & Gamble's push for mixed waste composting was met with opposition from both environmentalists and the waste industry, who were concerned that the finished product of mixed waste systems was environmentally hazardous and unmarketable (Holusha 1991). During the early 1990s, states began to impose stricter regulations on the end uses of mixed MSW compost, which added to the appeal of source separation (Goldstein and Steuteville 1994). The National Recycling Coalition, an alliance between industry and environmental advocates that had helped lead the design of curbside recycling systems (Lounsbury, Ventresca, and Hirsch 2003), pushed P&G to accept the growing consensus on source-separated composting (Raloff 1993). In 1993, the National Recycling Coalition held the National Source Separated Compost Symposium at a *BioCycle* conference in Minneapolis, Minnesota. The event included members of the Solid Waste Composting Council, the US EPA, the USDA, manufacturers, and environmental groups, who saw source separation as necessary for creating marketable end products and building the compost industry.

With the turn toward source separation, large manufacturers such as Procter and Gamble withdrew support for composting. This contributed to the relative absence of composting among the municipal recycling programs implemented in the 1990s (Norris-Waldt 2018).

---

policing of diaper waste should be understood in the context of a patriarchal waste regime, in which women are more likely to be tasked with managing waste, more likely to be sold wasteful products, and more likely to have their wastes scrutinized.

By the beginning of the current compost boom in 2005, source separation had become the norm for industrial organics recycling, and mixed MSW composting became a relatively marginal practice.[9] The model of municipal compost promoted by the Solid Waste Composting Council and the National Recycling Coalition mirrored curbside recycling collection: residents were responsible for putting organic waste on the curb, which would then be collected by municipalities, processed, and sold as compost. Source-separated curbside organics also carried the cultural attitudes of recycling: proper disposal is a civic duty that reduces municipal wastes in service of environmental and economic sustainability. A common refrain in user guides to new compost bins is that it's just like the other recycling programs that households are already used to (City of New York 2013b; SCDHEC 2018; DSNY n.d.).

CONCLUSION: LIMITATIONS AND POSSIBILITIES OF ORGANICS RECYCLING

While municipal organics recycling offers hope for a future that sends less waste to landfill than the present, existing programs in the United States have been shaped by a historically specific understanding of the goals and strategies of municipal organics recycling. High-volume diversion of waste is the primary goal. Centralized collection and large industrial systems are the preferred methods of accomplishing this, and these methods often produce a low-quality compost that is more likely to be used on highway embankments than organic farms (Gorrie 2015; Freeman, Skumatz, and D'Souza, 2012). Nonetheless, the moral frameworks provided by the organic and recycling movements portray organics recycling as progress toward closed-loop food systems and reduced ecological footprints for cities.

This chapter tells a familiar story of co-optation: while composting was promoted for social and environmental reasons, the predominant form of composting in the United States only marginally meets these ideals. The waste industry, manufacturers, municipalities, and legislators have forwarded a version of organics recycling that reflects the values of neoliberal waste management: building markets for collection, processing, and finished compost without disrupting the market for consumer goods.

9 Mixed waste composting is still used in rural areas where hauling is expensive, and tourist destinations where landfill space is unavailable and it would be difficult to train people on source separation (N. Goldstein 2005).

Organics recycling is progress, but it is progress toward a world where production and consumption remain fundamentally unchanged, and the waste industry will still stand to benefit from increases in waste production.

The standardized model of municipal organics provided the blueprints for New York City's curbside organics pilots, described in the next chapter. But community composters and microhaulers have their own ideas about what composting can be, and have their own approaches to organics recycling, built around different goals and values.

THE *BIOCYCLE* ARCHIVES have been a useful companion as I've researched the history of municipal compost systems. In its masthead, the magazine advertises itself as "the organics recycling authority." *BioCycle* occupies an important position that connects the organic origins of composting with the practices of organics recycling as a waste management strategy. It holds together many different ideas about what organics recycling is and can be.

BioCycle cover stories generally target industrial composters: usually they're about running an industrial compost or anaerobic digestion (AD) operation ("Facility Maintenance Strategies," "Feedstock Value of Compostable Packaging," "Catalyzing the Biogas Economy," and so on), and most of the advertisers are manufacturers of industrial composting equipment. When I first subscribed to *BioCycle*, I felt alienated by the apparent focus on industry—that's not what compost means *to me*.

However, every issue also features stories on community compost or the use of compost for environmental remediation: stories that reflect the wide array of possibilities that are contained within the concept of organics recycling. Here are some examples of stories from recent *BioCycle* issues:

- The nonprofit Givelove set up a composting toilet system at the Standing Rock protests against the Dakota Access Pipeline. This allowed protesters to hold their ground without creating undue strain on their environment (Johnston 2018).
- In the face of ongoing ecological devastation from farm runoff in Lake Erie, edge-of-field compost application might offer a means of slowing runoff (Zimnicki 2019).
- A compost training program in Washington, DC, teaches urban farmers how to remediate polluted soils and provide their farms with organic material (Bilsens Brolis 2018).

BioCycle editors have held space for an expansive concept of organics recycling. This reflects the organics recycling movement more broadly, which brings together industry, environmentalists, organic advocates, soil scientists, environmental justice activists, entrepreneurs, and community organizers—to promote "organics recycling."[10] What kind of organics recycling, in what context, for whom, by whom, is being debated. And that debate is shaped by power dynamics within the movement. Waste management has come to define the standard model of municipal organics, but the concept of "organics recycling" still holds together a variety of values and practices. This compost heap contains multitudes. Genealogy helps to demonstrate alternative possibilities in the history of organics recycling that can be repurposed for counterhegemonic composting futures.

Disposal is not necessarily the end goal of urban sanitation. Landfills and incineration are not necessarily the default solutions to the problem of excess stuff.

The practices and economic structures of farming could be more compatible with compost use. Compost farming does not need to be led by abstract values about sustainability; it could be pursued as a practical alternative to dependence on expensive, environmentally destructive fertilizers.

Industrial composting systems do not need to be built for waste management; they could be designed by farmers, landscapers, and gardeners who know what they want in a finished product. Some plants already describe themselves as "compost manufacturers" rather than "composters," emphasizing output over input (Alexander 1996). The compost system at McEnroe Farm, which I describe in the next chapter, has accepted municipal waste from New York City, but was built first and foremost as a way to create compost for the farm.

The relationship between "waste generators" and collection could be reimagined. Organics recycling could be remodeled after the community recycling centers of the 1960s and 1970s, which were understood as places where community interaction and jobs creation outweighed the goals of waste disposal (Lounsbury, Ventresca, and Hirsch 2003). The community composters and microhaulers I describe in chapters 3 and 4 have created systems that meet some of these goals. Alternately, mixed waste composting

10 This assortment of organics recycling advocates has more in common with Kim Fortun's concept of an "enunciatory community" (2001) than with most traditional accounts of social movements: in the welter of environmental politics, people come together across traditional lines of difference to articulate an idea—"organics recycling"—that emerges from the morass of the ongoing disaster of waste management.

could be the norm for organics recycling, which would obviate the need to train "waste generators" on source separation.

Organics recycling offers space for radical possibilities. In the context of neoliberal waste management, small-scale composting, compost manufacturing, and other alternatives might be at a disadvantage in comparison with industrial waste management approaches to organics recycling. But the concept still includes alternative systems in both rhetorical and practical ways. The institutionalization of the industrial approach has standardized a form of organics recycling that fails to live up to the dreams of many compost promoters, but it does not mean the loss of community-oriented, just sustainability-minded composting systems. Rather, these approaches coinhabit the category of "organics recycling" in complex relationship. In the following chapters, I describe how actors in municipal sanitation, community composting, and microhauling differ in how they understand the goals and practices of organics recycling, while still collaborating on the larger project of making New York City compost.

CHAPTER TWO

The NYC Curbside Organics Collection Pilot

BUILDING THE MARKET FOR ORGANICS RECYCLING

In fall of 2013, the New York City Department of Sanitation (DSNY) started collecting food scraps and other organic wastes from 32,900 households and 350 public schools in a pilot program that was intended to test the feasibility of curbside organics collection in New York City. The pilot began in eleven neighborhoods in the Bronx, Brooklyn, and Staten Island. DSNY asked residents to separate out food scraps, soiled paper, and yard waste from the rest of their garbage and put it in specially designed brown plastic bins on the curb on specified days. In October 2015, a DSNY report said that the curbside organics pilot had been expanded to serve 500,000 residents and 700 schools, and was confident that the pilot had "demonstrated that New Yorkers are ready and are willing to take part in this important civic undertaking" (DSNY 2015a, 47). The 2015 City Plan, OneNYC, promised that the city would offer organics collection in all 59 community boards by 2018 (Karidis 2016).1

DSNY halted expansion of the pilot in May 2018. At that point, it was offering curbside organics collection to 3.5 million residents in 24 community boards—40% of the city (Amiri 2018). But the program had minimal impact on the city's diversion rate: only about 11% of the compostable material in the program areas was being collected (Champeny 2018). Experts and administrators provide various explanations for the low diversion rate: lack of education, lack of incentives, or the unfamiliarity of separating out

1 When DSNY employees would talk about this goal during my fieldwork between 2014 and 2017, they often made a specific expression, where their eyes go big and they stare at something just past my shoulder. There was disbelief in this expression, and fear, but also hope.

organics (Amiri 2018; Muoio Dunn 2018; DSNY 2017). Between 2018 and 2020, DSNY, the mayor, and city council explored strategies for increasing diversion rates, but the program was put on pause in 2020 due to budget cuts in the face of the COVID-19 pandemic (Rubinstein and Mays 2020). Mayor Bill de Blasio announced his plan to reinstate a modified, opt-in organics program in his 2021 Earth Day press conference. Despite the city's stated intent to build a zero-waste future (City of New York 2015), its efforts to institute curbside composting have faltered, and the future of municipal organics recycling in New York City remains uncertain.

In this chapter, I describe how the rise of municipal organics recycling has been playing out in New York City.[2] This story starts with the organics recycling pilots in the 1990s and early 2000s, and their role in creating a network of compost advocates and educators. I then describe the beginnings of the current pilots, and how the city has pulled together collection routes, waste habits, processing infrastructure, cultural values, and the promise of anaerobic digestion in order to craft an infrastructure for composting. This story ends with the cancellation of curbside organics collection at the beginning of the COVID-19 pandemic, and the role of community composters in the #SaveOurCompost campaign that fought for the reinstatement of curbside organics. The chapter concludes with a discussion of the New York City curbside organics program in the context of neoliberal waste management, as well as the meanings it holds for community composters and microhaulers.

Many policymakers, city employees, and activists worked to meet sustainability and environmental justice goals with curbside organics in New York City, and I do not want to dismiss their efforts. However, the pilot that began in 2013 was limited by DSNY's institutional charge to empty the streets of waste, and the dominant logic of neoliberal waste management in which they worked toward that goal. Modeled after the dominant approach to organics recycling described in chapter 1, the pilots were oriented toward cutting costs, building markets, and selling New York City as a sustainable city. But activists have found new possibilities in curbside compost, and have advocated for alternative approaches that would prioritize environmental justice, jobs creation, and local ecosystems.

2 This chapter focuses on the residential organics collection pilots, but this has not been the only large-scale composting system to address waste problems in NYC. Large-scale commercial haulers have been collecting organics from some commercial and industrial waste generators in NYC for years, but they will play only a minor role in this account.

THE EARLY DAYS OF MUNICIPAL COMPOSTING

The New York City Department of Sanitation's early organics recycling pilots were part of the national turn toward resource recovery in the late 1980s and early 1990s, shaped by garbage scares, federal legislation promoting diversion, and the emerging recycling industry (Lounsbury, Ventresca, and Hirsch 2003; MacBride 2011). They also responded to local conditions: in 1984, mounting disposal concerns prompted mayor Ed Koch to propose five new incinerators across the city. In response, environmental groups and neighborhood advocates demanded less polluting, more environmentally just forms of disposal (E. Goldstein 2014). New York City was under additional pressure from the New York State Solid Waste Management Act of 1988, which pushed the city to design an integrated solid waste management plan that would account for 50% waste reduction and 50% waste-to-energy (J. Clarke 1993).

Local Law 19 of 1989 created the recycling systems for households and institutions that are in place today. While the law focused on nonputrescible recyclables (metal, glass, plastic, and paper), it encouraged expansion into other materials, including organics (J. Clarke 1993).

Starting in 1991, DSNY asked residents of Park Slope to separate out food scraps from their trash for collection. The program provided educational materials, support for backyard and community composting, and a household hazardous waste pickup day. Collected materials were trucked to a newly constructed composting facility at the Fresh Kills Landfill on Staten Island. The pilot was relatively successful: by the second year, it was diverting 41% of the area's food waste (Lange 2001; J. Clarke 1993).

Park Slope was an obvious choice for the first organics pilot because it was a middle- to high-income neighborhood with a history of environmental activism and plenty of space at curbside for an extra bin. In 1993, the department began collecting source-separated organics from Starrett City, a federally assisted housing development in eastern Brooklyn. Encouraged by the Park Slope pilot, DSNY wanted to explore organics diversion in high-rises, which offer unique challenges for collection. Residents of Starrett City were also low-income and predominantly Black.

DSNY may have seen race as an important variable for a compost pilot. The inaccurate and racist idea that Black people are worse at recycling than white people was, and continues to be, influential in conversations around waste, and has affected the deployment of recycling resources in communities of color (DNLee 2011). Despite evidence that there are no differences in environmental attitudes between whites and people of color (Parker and McDonough 1999; Mohai 2003), I've seen many white environmentalists at public meetings ask veiled (or unveiled) questions about how to get people of color to recycle and compost. Activists and scholars have suggested that statistics showing lower recycling rates in predominantly Black neighborhoods are either inaccurate or reflect the failure of recycling programs rather than their communities (DNLee 2016; Mock 2014; MacBride 2011).[3] Bhahij Chancey points out that the 2010s organics pilots were more likely to serve white residents than Black, Asian, or Hispanic residents (2021).

Starrett City, unlike many high-rises, had utility rooms on every floor and space for dumpsters, which DSNY assumed would make it an easier place to implement composting. However, odors and the threat of vermin made residents wary about leaving their food scraps out in the utility room for up to a week. Many residents refused to separate out their scraps, and others wrapped their scraps in enough plastic to create a contamination problem. A DSNY report frames this as a failure to educate residents (Lange 2001), but DSNY also failed to understand residents' needs.

Neither the Park Slope nor the Starrett City pilot performed well enough for DSNY to expand residential organics in the 1990s. DSNY concluded that, without incentives for separation, collecting organics wasn't going to save money or have a significant environmental impact (Lange 2001).

DSNY continued its education programs after the pilots folded. In 1993, DSNY's compost education was institutionalized into the NYC Compost Project, which has provided outreach, training, and support for backyard and community composters. One legacy of the early pilots is a sense that the "ick factor" of source separation was going to be a major hurdle in creating an organics collection program for the city (Walters 2018). The Compost Project has been one of the department's tools for addressing the "ick," and cultivating a "culture of composting" among New Yorkers.

The recycling law also created the Citywide Recycling Advisory Board (CRAB) and a Solid Waste Advisory Board (SWAB) for each borough. The CRAB and SWABs were all-volunteer organizations tasked with reviewing

3 This resonates with my own experiences as a white person trying to recycle in the predominantly Black, middle-class neighborhood of Prospect Lefferts Gardens. Local bodegas and supermarkets did not always sell the clear bags that DSNY requires for recyclables; I had to leave the neighborhood to find them. Even if curbside collection is available to all New Yorkers, the infrastructures for recycling are not equally accessible.

the new recycling plan, and they became platforms for waste reduction advocacy. In 1992, the CRAB and the Natural Resources Defense Council (NRDC) successfully sued DSNY for failing to meet its diversion goals, forcing DSNY to build up its recycling program. The same year, the CRAB released its report "Recycle First," which advocated for organics recycling along with other waste reduction goals and serious reconsideration of the city's incineration plans (J. Clarke 1993).[4] The Manhattan, Brooklyn, Bronx, and Queens SWABs continue to meet and play a role in waste advocacy.

Soon after the curbside pilots, DSNY pursued neighborhood-scale composting in enclosed digesters, which would speed up decomposition and trap odors. A 1994 report describes an enclosed composting system being installed at the Brooklyn Botanic Garden and a similar planned system at Rikers Island Prison Complex.[5] While the Rikers Island digester remains operational, in-vessel composting did not expand beyond a few projects.

DSNY also seriously explored backyard composting as a solution to waste problems created by the closure of the Fresh Kills Landfill. Fresh Kills had been the largest landfill in the world for decades, and was the primary destination for New York City's waste. In 1996, in response to activist pressures,[6] the state of New York ordered Fresh Kills closed within five years. While this was a boon for Staten Island, it meant that New York City needed a new destination for its waste. A task force appointed by mayor Rudy Giuliani and governor George Pataki pointed to backyard composting as a partial solution. Starting in 1998, DSNY partnered with the four Borough Botanical Gardens on a Backyard Composting Pilot Program. The Botanical Gardens conducted extensive solicitation and outreach to give residents access to compost bins and training on how to use them (Lange 1999).

In waste audits, the Botanical Gardens found significant reductions in food waste among the households that volunteered, and even among the

4 The report also recommended a four-bin recycling system (paper; organics; metals, glass, and plastic; and landfill) that could recycle the majority of the city's waste by 2001, and the creation of a waste prevention office within DSNY dedicated to building recycling infrastructure and reducing the use of packaging and single-use items in NYC.

5 The composting facilities on Rikers Island became operational in 1998, and continued to function through 2021. While city officials have celebrated the facilities as part of a work-training program for inmates and a more sustainable Rikers, the composting effort should also be understood as part of the politics of the prison complex. In part, the on-island composting system is meant to reduce the facilities' reliance on waste services—which might offer escape routes. And as a *New York Times* piece quips, Rikers also offered a place where neighbors couldn't complain about unwanted waste infrastructure: "Nobody had to ask the inmates for their view" (Martin 1999).

6 In part pressure from the Staten Island SWAB.

neighbors of participating households. But the low rate of participation and the relative rarity of backyards in New York City meant that backyard composting probably couldn't make a meaningful dent in the city's organic waste (Lange 1999).[7] Still, DSNY and the Botanical Gardens continued backyard composting education because it "promotes awareness of where waste goes and how it is managed" (11). Backyard composting education and discounted compost bins have been folded into the Compost Project.

COMPOST FOR DEVELOPMENT UNDER MAYOR BLOOMBERG

Compost gained a new currency during the tenure of mayor Michael Bloomberg (2002–13), who oversaw planning initiatives and pet projects that set the terms for composting and food waste initiatives in New York City. Much of this progress was a result of activists pushing for more responsible waste management. But Bloomberg-led projects—including PlaNYC, the Mayor's Food Waste Challenge, the commercial organics ban, and the curbside organics collection pilots—linked composting to development and market creation (Checker 2011).

In 2006, Bloomberg tasked the newly formed Office of Long-Term Planning and Sustainability with creating a sustainability plan for the city. When it was released in 2007, its PlaNYC was greeted as "bold and ambitious" (Shepard, Tyree, and Corbin-Mark 2008) and praised for its responsiveness to environmental justice concerns (Rosan 2012).[8] The Mayor's Office has released an updated PlaNYC each year from 2007 to 2014, after which the plan was renamed OneNYC. Since 2011, PlaNYC has included organics recycling as part of a toolkit for reaching 75% waste diversion by 2030.[9]

While PlaNYC was revolutionary for putting urban planning in conversation with climate change, critics have argued that Bloomberg's version of sustainability is dedicated to development and incompatible with equity (Checker 2011; Agnotti 2008; Rogers 2015). Heather Rogers describes PlaNYC as greenwashing, putting a sustainable front on the massive

7 The report includes a handful of imaginative figures: if the city achieved a 60% participation rate in qualifying households, and if they doubled weekly diversion, the backyard composting pilots might achieve a 2% diversion rate. But it reminds the reader that this is likely an overstatement of the possibilities of backyard composting

8 The advisory board for PlaNYC included two environmental justice organizations: the West Harlem Environmental Action Coalition and UPROSE.

9 In PlaNYC, organics recycling here includes community composting, residential organics collection, commercial organics collection, anaerobic digestion, and others.

development taking place in New York City, while exporting the harmful effects of manufacturing and consumption elsewhere (2015). Melissa Checker describes PlaNYC as "environmental gentrification": replacing industrial properties with less polluting housing and commercial development, thereby pricing neighbors out of their houses and reducing job opportunities, and forcing industry elsewhere. While PlaNYC envisions a "greener, greater New York City," these benefits accrue only to those who can afford to live, vacation, or own property there, and displaces the environmental impacts of development elsewhere (2011).

The Mayor's Food Waste Challenge, announced in April 2013, got more than 100 restaurants to pledge to divert at least 50% of their food waste from the landfill "through composting and other waste prevention strategies" (City of New York 2013a).[10] The challenge offered regular meetings to discuss strategies, technical advice on reducing waste, and promotion on the nyc.gov website as well as the opportunity to self-promote as a participant in the challenge. Restaurant food waste consultants argued that "it's simply good business" to reduce food waste (Integrated Waste Management Board 1992). Many of the restaurants that signed on to the challenge were already separating out their food scraps in order to save money and meet their own sustainability goals, but joined the challenge in order to show support for reducing food waste (Forgrieve 2014). In the first six months, 50 of the restaurants that signed on to the project met their 50% reduction goal, and diverted 2,500 tons of food waste.[11] Katrin Bennhold characterizes this kind of policy as the "nudge approach," which creates the appearance of government action but "works to shrink government services and replace traditional public policy" (2013).

In a similar vein, the commercial waste ban offered regulation rather than support. In September 2013, Bloomberg worked with Leititia James, then chair of the Sanitation Committee of City Council, to develop legislation that would require food manufacturers, wholesalers, retailers, and food service establishments in hotels and arenas of a certain size to separate out food waste and contract with haulers to recycle it. The committee report states that the proposed legislation is intended to create a reliable source of organic material that can make the construction of a composting facility

10 While the press release includes donating uneaten food to shelters and food banks, Bloomberg had previously tightened regulations on donating food, citing concerns about its nutritional value.

11 The most recent estimate of commercial food waste I can find is 234,000 tons/year (HDR 2004).

economically viable, avoiding the chicken-and-egg problem of building a facility first versus generating a stream of organic waste: "This lack of participation appears to have a cyclical effect, thereby limiting the establishment of a reliable stream of material. One possible way to address this disconnect is to establish a requirement ensuring that there will be a steady and reliable stream of source-separated organics in the future" (James 2013, 3). The commercial sector was an appealing place to pursue organics collection in New York City because it's entirely served by private carters, not DSNY, and so would require minimal funding from the city.

At hearings, the proposed regulation drew support from composting and anaerobic digestion professionals, who agreed that a "reliable stream" of organics could make their businesses sustainable. It also drew support from large environmental organizations such as NRDC and smaller local environmental justice groups, and from the Manhattan and Brooklyn SWABs. However, it saw pushback from carters (because it would require new and costly hauling equipment) and from restaurant associations, who claimed it would force an added cost on small restaurants that were already struggling. The bill was passed into law at the end of 2013, to take effect in 2015. DSNY commissioner Kathryn Garcia postponed the start date until 2016 and restricted the ban to affect only larger establishments (E. Goldstein 2015).

The curbside organics collection pilot was Mayor Bloomberg's final contribution to compost in New York City. In his State of the City address in February 2013, he announced the beginning of the curbside organics pilot, proclaiming food waste to be "New York City's final recycling frontier." He announced the pilot curbside program, and explained organics recycling in the language of economic efficiency: "We bury 1.2 million tons of food waste in landfills every year at a cost of nearly $80 per ton. That waste can be used as fertilizer or converted to energy at a much lower price. That's good for the environment and for taxpayers" (Navarro 2013).

The development values in PlaNYC are part of a tradition that Miriam Greenberg traces back to the city's "comeback" in the late 1970s. Greenberg describes how city leaders responded to the fiscal crisis and image crisis of the 1960s and 1970s by "branding" New York City as a culturally interesting, business-friendly, safe destination for tourism, consumption, and investment. The image shift was accompanied by policy shifts, and New York City leaders divested from racialized neighborhoods, laid off municipal workers, and offered tax breaks to corporations (2008). Greenberg argues that Bloomberg's—and later de Blasio's—cultivation of a sustainable image for New York City should be understood as part of an ongoing effort to sell the city to investors and businesses as a cutting-edge, climate-proof metropolis, rather than simply an effort to build a sustainable city for the people who live there (2015).

THE CURBSIDE ORGANICS COLLECTION PILOTS

The Department of Sanitation started a new curbside organics pilot in May 2013, collecting food scraps and yard waste from 3,200 households in the Westerleigh neighborhood of Staten Island. The Westerleigh pilot was met with general support from residents, and achieved a 43% participation rate within its first month (Navarro 2013). In June, council member Deborah L. Rose of Staten Island introduced legislation that would guarantee the continuation and expansion of the pilot. The bill set parameters for a two-year pilot program with regular reports to the mayor and city council on its diversion rates, and regular expansion into new areas.

At a Sanitation Committee hearing that month, sixteen New Yorkers—mostly community composters and permaculturists, but also SWAB members, DSNY employees, a teacher, and an NRDC lawyer—turned out to express their support for the proposed pilot program. The community composters tempered their support with concerns about some of the details left out of the bill: given the lack of processing facilities, materials might wind up in the wastewater treatment plants, where it would be digested into methane and sewage sludge ("or biosolids, as it is now politely re-termed" added one community composter, sardonically).[12] The bill made no provision for local composting, and did nothing to support the community compost systems already in place. And while a compost program promised to create more jobs, the activists needed some assurance that these would be good jobs, and not just more exploitive waste industry jobs. Finally, they demanded that the pilot be committed to waste equity—not just in terms of processing and transporting organic waste but also in terms of collecting it.

12 Biosolids split the rhetorical and technological field of organic waste. While the sludges created by AD facilities have been used to fertilize food crops, many activists (often proponents of local systems) protest that the materials are unsafe, that this is a matter of water treatment pawning its wastes off onto farmers. The recent identification of PFOAs in sewage sludge fertilizer has underscored the danger of closed-loop systems (Treat 2021).

The bill was approved by the city council in September, 44 to 1, and became Local Law 77.[13]

DSNY expanded the pilot to nine more neighborhoods in the Bronx, Brooklyn, and Staten Island in September and October 2013. Its selection criteria for new neighborhoods included a history of high recycling rates and low population density (DSNY 2015a). Park Slope and Windsor Terrace in Brooklyn were high-income, predominantly white neighborhoods with a history of environmental activism. Edgewater Park and Country Club in the Bronx were predominantly white, middle-income neighborhoods. The neighborhoods in Staten Island were predominantly low-income and mixed race.

Because high-rises pose a separate set of problems from single-family dwellings and small apartment buildings, DSNY didn't implement any collection districts in the crowded borough of Manhattan. Instead, it offered pickup to buildings of 10 units or more that volunteered and organized in order to receive organics collection.[14] The Durst Organization, one of the oldest family-run real estate companies in New York City, offered up the Helena, a 600-unit luxury high-rise on West 57th Street in Manhattan, as a testing ground (DSNY 2015a). The Helena pilot also established a relationship between DSNY and McEnroe Compost, a facility that has remained important for the city's organics program.

The curbside organics pilot served 32,900 households in fall 2013. By fall 2015 DSNY offered organics collection to 190,500 households in 41 neighborhoods, with 69% of these residents participating.[15] At this point, DSNY estimated that the organics program was capturing 24% of the compostable material from participating households, though capture rates varied by neighborhood, and were higher in less dense areas. By the time expansion was halted in 2018, 3.5 million residents had access to curbside organics. In 2018, city data show the food waste diversion rate for participating neighborhoods

13 The sole dissenter was Diana Reyna of District 34 (Williamsburg and Bushwick, two historically overburdened low-income neighborhoods of color), who had in the past proposed legislation that would mitigate the burden of transfer facilities on overburdened neighborhoods.

14 Because high-rises were a problem in previous pilots, the department has treated these buildings as special cases, and has devoted an ethnographic research team, headed by Samantha MacBride, to try to understand the way that waste practices change in these buildings. MacBride points to the importance of networks for changing habits. Mostly, she argues, these are luxury buildings that embrace organics collection as part of the green building movement, or mid- to high-end co-ops that are prime sites for tenant organizing (2015a).

15 Despite the high participation rates, diversion rates have been very low. The 69% participation rate includes everyone who sets out compost more than once a week. Many residents are likely not separating out all possible food scraps.

dropped to 10.7%. This was caused mostly by the addition of areas with low diversion rates, but also by reduced diversion rates in some areas (DSNY 2023). Over the first five years of the pilot, DSNY collected 83,000 tons of source-separated organic waste from households, about 1.5% of the household organic waste generated in New York City during that period.

Mayor Bill de Blasio took office in 2014, inheriting the curbside organics collection pilot from his predecessor. De Blasio embraced curbside organics, and wrapped the pilot into his own sustainability plan, OneNYC. One of the most visible promises of the new plan was making New York City into a zero-waste city by 2030 (shortened to "0X30" in advertisements and announcements). The waste reduction plan included a bag tax, beefed-up recycling programs, and a "save-as-you-throw" policy that would reward residents for disposing of less trash (Jorgensen 2015). De Blasio oversaw the growth, cancellation, and eventual rebirth of the curbside organics program.

EMBRACING THE BROWN BIN

The organics pilot required a high level of engagement from participants. In order to make the organics pilot work, DSNY and the Mayor's Office needed to create a physical and cultural infrastructure that could support the habits of source separation. The brown curbside bins played a central role in this campaign, and the city worked to demonstrate that the bins could be sanitary and environmentally friendly parts of residents' daily lives.

Worried that residents would feel uneasy about attracting rodents by putting bags of concentrated food scraps on the street, DSNY put out a request for proposals for plastic bins that could latch tightly to keep critters out. The resulting bins were built out of tougher plastic than average garbage toters, and I never saw them develop the holes that rats and raccoons like to dig in the corners of bins.

Raccoon resistance was a selling point for the brown bins. In a video posted to the DSNY Facebook page in March 2019, a human in a raccoon costume fumbles with a brown bin, fails to open it, and walks away frustrated. A voiceover encourages residents, "Separate your food scraps to help stop feeding the city's wild residents." The organizers of the curbside organics program argued that the bins create a positive public presence for curbside collection: "Our best educational and operational tool will continue to be the brown bin itself. The bin conveys a powerful message of safety, organization and commitment on the part of the City [to] help residents comply with organics recycling" (DSNY 2015a, 47–48).

In their study of organics recycling in London, Alan Metcalfe and colleagues suggest that food waste bins themselves play an important role in changing the habits of consumers (2013). In their analysis, "changing hearts and minds" (149) of waste generators—convincing people to care about composting—isn't as important as creating infrastructures that can enter a household without disruption.

Mayor Bill de Blasio and his family tried to convince New Yorkers to allow the bins into their home in the one-minute video "#BrownBin: The de Blasio Family on Food Waste and Recycling." The video was uploaded to YouTube on August 6, 2014. The video had 24,989 views as of July 11, 2021, with 63 likes and 3 dislikes. It was produced and released by the Mayor's Office.

Mayor de Blasio's daughter, Chiara, begins the video by explaining that "recycling food and yard waste is a lot easier than people think." Her explanation is played over footage of her transferring the contents of a kitchen countertop bin (which appears to be full of salad) to the brown bin at the curbside of her family's Park Slope house. Food waste, she says, is just as easy to take care of as garbage or plastic/paper/metals recycling. Lighthearted music featuring a xylophone plays in the background.

Cut to the de Blasios' upper-middle-class kitchen. It's clean and spacious but not showy. Bill talks to Chiara while she and her brother, Dante, eat watermelon. Bill's wife, Chirlane McCray, fries eggs in the background. Bill is white, Chirlane is Black, and Chiara and Dante are biracial. Dante carries the tan kitchen counter collector over to the table and opens it for his dad, creating an ersatz basketball hoop into which the mayor stiffly tosses a watermelon rind, his fingers snapping back into a loose fist that lingers in the air for a second as he makes sure his toss was successful. The motion is arch nerd, and the full awkwardness of de Blasio's white dad identity is on display here. While I imagine that it was used to communicate a "dad bonding with son over compost" moment, it also provides a symbolic bridge between the "nerdy white hippie" image of composting and the Mayor's Office impression of an "average Black youth."

Bringing the video back to its beginning, we follow Dante as he takes the countertop collector and dumps it in the brown curbside bin in the backyard, then pulls the curbside bin out to the street. Dante is tall, and he hunches as he interacts with the bin, which we now see is comically small compared to a trash can. After Dante puts the bin on the street, Chiara closes the video by explaining that not everyone has a brown bin yet, so if you don't have it, you can just take your food waste to a local food waste drop-off center.[16]

Messaging around the bins explained how easy it would be to integrate composting into a happy, hygienic household. But with an object as unstable as rotting food, rhetorical containment can prove challenging. Despite DSNY's best efforts to make putrescible waste feel less putrescent, some people in pilot areas questioned the sanitariness of curbside organics. In a survey conducted in June and July 2015, 35% of residents who did not participate in the pilots reported concern about rodents or bugs as one of their reasons, and 35% listed odor (DSNY 2015a).

Worries about sanitariness precipitated into public discourse in Greenpoint in the summer of 2015, when organics collection extended into the North Brooklyn neighborhood. Greenpoint has a long history of housing disproportionate amounts of waste infrastructure, including the Newtown Creek Wastewater Treatment Plant, waste collection routes, and material transfer facilities. And communities in Greenpoint have a long history of fighting against these infrastructures (Yardley 1998; Sengupta 1996; M. Moss 1991). Shortly after food scrap collection was implemented in June 2015, news stories started running about Greenpoint residents "making a stink" about the odor and insect problems associated with the new pilot projects: "Area residents have reportedly been bemoaning the stinky compost bins all week, with locals complaining to various news sites that the rotting food was attracting maggots, flies and roaches, in addition to emitting an unpleasant odor. 'It smells like dirty diapers and vomit all up and down the block, and the smell lingers for days,' one resident told CBS 2" (Fishbein 2015). The Department of Sanitation dismissed residents' concerns, saying that as with any new program, Greenpoint was on a "learning curve" with organics recycling, and residents were simply working out the kinks in separating out their scraps (Furfaro 2015).

THE BACK END

As the activists commenting on Local Law 77 had predicted, the Department of Sanitation didn't have a clear vision for what to do with the organic

16 Likely a greenmarket or a neighborhood garden.

waste that it collected. Processing capacity became an issue as the pilot expanded. While the materials from the Westerleigh pilot could be processed locally at the Fresh Kills Compost Facility, the department floundered to find capacity for the rest of the city's organic waste.

The capacity problem wasn't new. Before the organics pilots, DSNY had struggled to find space to compost yard waste. Faced with the cost and truck traffic involved in carting all of the city's leaves to the composting facility at Fresh Kills, DSNY partnered with the Parks Department to compost leaves on Parks land and to use the finished product to restore soil in city parks (Lange 2001). However, this symbiosis could only go so far with food waste, which smells unpleasant and attracts rodents.

In 2012, DSNY posted a request for proposals for contractors who could "furnish all supervision, labor, material, and equipment necessary to take delivery of, process, and convert to beneficial use a minimum of 20 tons per day (TPD) of source-separated food waste and other organic waste" (A1–1). In October 2013, DSNY awarded the contract to four companies: Waste Management ($1.4 million), Regal Recycling ($800,000), Brooklyn Transfer ($525,000) and WeCare Organics ($250,000) for a total of $2.9 million on tipping fees for the organics pilot. Food waste from the pilot was shipped to transfer stations in North Brooklyn, Harlem, and Jamaica, Queens, where it was picked over and exported to sites ranging from Andover, New Jersey (60 miles from New York City) to Marlborough, Massachusetts (193 miles from New York City).

Waste Management and WeCare both carted portions of their waste 123 miles south to the Peninsula Compost Facility in Williamsburg, Delaware. The facility was designed to process 120,000 tons of food waste per year (330 tons/day) using statically aerated windrows. Aerial photos of the Peninsula facility before it shut down suggest a mechanical orderliness—an asphalt pad with 54 neat, covered windrows packed together, with buried channels leading to a leachate pond, all squeezed into an industrial landscape between rail tracks, roads, and power lines. At the time of its construction, Peninsula was the largest compost facility in North America (Guevarra 2011). However, the facility ran into trouble soon after it opened. In 2012, a fire sent up enough smoke to close down the Port of Wilmington.[17] Safety inspections revealed enormous piles of mixed waste not yet being

17 Fires are common at compost facilities, and are most commonly caused by spontaneous combustion of rapidly decomposing organic materials (Coker 2021).

composted, as well as contaminants that had been picked out and left in heaps, which resulted in fines from the Department of Natural Resources and Environmental Control. Offensive smells in disadvantaged areas were an environmental justice issue, argued Delaware General Assembly representative James Johnson (Montgomery 2014), and when Peninsula asked to renew its permit in 2014, residents turned out to complain about the odor. The permit was not renewed, and the Delaware Department of Natural Resources and Environmental Control ordered the facility shut down in late 2014 (*BioCycle* 2014).

The waste stream coming to Peninsula was poorly screened, and the waste it brought in was highly contaminated. When I spoke to Erich McEnroe, the facilities operator at McEnroe Compost, he explained that the refuse stream coming to Peninsula from Waste Management and WeCare was far below the standards of McEnroe.

G: Do you know if you wound up getting more [materials] with the shutdown of Peninsula?

E: A couple haulers contacted us, we took one look at one of their loads, and we turned it away.

G: Wow!

E: There's a reason why Peninsula got shut down.

Furthermore, the facility brought in a high ratio of greens to browns (120,000 tons of food waste to 40,000 tons of leaf and yard waste), which promotes anaerobic activity—the primary source of odors in composting.

When Peninsula closed, it left DSNY scrambling to manage thousands of tons of organic waste. The closing of Peninsula was still a fresh wound when I started attending public meetings about compost. Mentioning Peninsula to DSNY employees usually elicited a sigh.

McEnroe Compost offers a much more optimistic image of what the back end can look like. McEnroe Farm is two hours north of the City in Millerton, New York. McEnroe received about 2,500 cubic yards of food scraps per year (about 3 tons/day) from New York City.[18] The composting facility at McEnroe started in 1989, when Douglas Durst—chairman of the Durst Organization—entered a partnership with Ray McEnroe to compost the

18 The incongruity between DSNY's weight measures and McEnroe's volume measures results from the permitting process for compost facilities, which are allowed to compost a certain volume of food waste per year. Waste haulers, however, measure their materials by weight, likely because they rely so heavily on compacting technologies.

bedding from his horse farm on a small section of the McEnroe property. The project started accepting outside feedstock in 1990, and began taking food scraps from New York City in 2008, through a composting program at the Durst-owned Bank of America Building.[19] McEnroe started accepting materials from the Durst Organization's Helena Building at the beginning of New York City's organics collection efforts in 2013.

When I visited McEnroe in 2015, it had just replaced its windrow turner, and its windrows were steeply stacked about eight feet high. "That's what I dream about at night," Erich half-joked when I admired them. The composting operation blended into the landscape—it was situated in a clearing on a hillside, with just a small dip in the ground separating the three acres of compost from the surrounding field. The turner was ambling down the compost heaps, mixing and grinding and kicking up huge clouds of steam. For the most part, the smell was just the combination of hay and rot and healthy soil that you get at any farm, a sort of background fecundity that suggested life and growth more than death and decay. Every once in a while, the machine would hit a patch of something that was rotting wrong, and the smell would travel 200 feet up the hill to smack us like a water balloon, so that we'd make faces and laugh. Andrea, McEnroe's compost educator, explained that these were probably little pits of manure.

The area around McEnroe is not as busy as Wilmington, but it's not empty. The roasting facility for Irving Farm, a hip coffee retailer, is just a quarter mile from McEnroe's compost facility, and a handful of other households and farms are located in the valley. Erich pointed at the clouds of steam from the windrow turner, climbing up the hill toward us, and explained that they try to keep an eye on the wind and make sure they turn the compost only when it's blowing uphill, away from the neighbors. McEnroe has butted heads with its neighbors in the 30 years they've been composting. But they've served a function in the community as a supplier of compost to other small farms. Occasional odors have not significantly detracted from these relationships.

While McEnroe Compost is a commercial waste management operation—it charges $35 a ton to accept organic waste—it is also a farm and a

19 The Durst Organization's involvement in compost is part of a broader investment in "green development." In 1990, the organization funded the construction of 4 Times Square, the nation's first large-scale office building to incorporate green design principles. Douglas Durst, CEO, explained in an interview that "if you're not building a green building, you're building an obsolete building" (Travers 2010). The Durst Organization has received awards from the NRDC, the EPA, and Global Green for leading the charge on green building (Storey 2005).

purveyor of quality compost and soil. A 40-pound bag of McEnroe compost sells for about $8.65 at a garden store. As Erich explains, "We like to have a tipping fee, it's nice, it keeps us operational, but our concern is to have a good finished product." McEnroe Farms, unlike many of the other compost facilities in the New York City waste system, was created with soil in mind, rather than waste management. The "closed loop" vision, in which food waste is processed into compost in order to grow more food, seems very close to the McEnroe reality: it uses 60–70% of its finished compost in its own operations. This might explain why McEnroe is often held up as an exemplar of the New York City organics system.

Most of New York City's organic waste isn't going to farms like McEnroe; it's going to facilities that were built as waste processing facilities. New Milford Farms in Connecticut is a facility built to compost waste from a nearby Nestle plant (Gross et al. 1997), and is now predominantly owned by Waste Management. The Bedminster-Marlborough facility, operated by WeCare Organics, has been sued by the city of Marlborough for odors (Hatch 2013). WeCare has also been sued by the city of Rochester (Orr 2014) for dumping organics in a landfill when the city had paid to compost them. When composting is framed as waste disposal, it's easier for economic pressures to encourage bad management practices.

These two different styles of doing compost demonstrate the possibilities within the concept of organics recycling: while Peninsula (like many of the other compost facilities receiving New York City's waste) was designed by waste managers for the purpose of disposal, McEnroe was designed by farmers in order to create a soil amendment. McEnroe manages the waste it takes in and the product it puts out in a metabolism with the farm and the environment around it. Industrial facilities like Peninsula have no such relationship to the land or their surrounding communities.

While the phrase "conver[sion] to beneficial reuse" offers space for both McEnroes and Peninsulas to become part of the city's organics recycling infrastructure, the economics of waste management favor Peninsulas. Without intentionally pursuing compost production over disposal, increasing demand for processing capacity will encourage investment in facilities that are capable of accepting higher volumes and are less demanding about the quality of inputs. McEnroes are possible in this system, but they're not the norm.

ANAEROBIC DIGESTION

Anaerobic digestion (AD) is a process in which organic material is decomposed in the absence of oxygen to produce biogas: a mixture of methane and carbon dioxide that can function as a substitute for natural gas. Many cities, including New York City, already use AD to stabilize sewage sludge,[20] but the methane generated by digestion is not usually enough to power the process. Food waste has much more potential to produce methane than sewage, so codigestion with sewage sludge could allow sewage plants to power themselves and potentially generate excess methane that could be used for other purposes (Banks et al. 2018).

AD had been on the table as one of the city's organics recycling options since the early 2000s, but the city moved toward composting for a variety of reasons. AD requires expensive systems that can capture escaping gases, as well as further processing of digested materials. Composting is less demanding in terms of up-front capital investments, and the technology for composting has been tested in a way that AD hasn't. Furthermore, the uncertain status of AD among environmentalists (a member of the Brooklyn SWAB called the process "waste to sludge" at the 2013 hearings) made it a politically iffy choice, even though it promises a more fungible return (natural gas) than composting.

Nonetheless, AD has remained within the realm of possibilities for organics recycling in New York City.[21] One proposed project involved a partnership between the Department of Environmental Protection (DEP), Waste Management, and National Grid. Representatives of each organization, along with Bridget Anderson from DSNY, explained the plan at a public meeting in Williamsburg in June 2015.

In the project, food scraps from schools and businesses were delivered by truck to a Waste Management-owned facility in Greenpoint, where they were ground into a slurry using a patented technology called Centralized Organic Recycling equipment (CORe). "It looks like oatmeal but it sure doesn't smell like oatmeal," joked Jim Van Woert, the representative from Waste

20 Stabilization is the process of reducing the odor and activity of sewage sludge. It makes it possible to transport sludge for disposal elsewhere.

21 The 2006 Solid Waste Management Plan included a call for an AD facility to be built to process waste at the Hunts Point Food Distribution Facility in the Bronx (DSNY 2006). A feasibility study in 2010 suggested that such a facility could be built and operated profitably (R. W. Beck 2010), but the DEC never pursued construction, likely due to budgetary concerns. In 2012, Mayor Bloomberg opened an RFP for a "clean" waste-to-energy plant within 80 miles of the city, and included AD as a possible technology, but it got no takers, likely because there was not yet any guarantee of a supply of source-separated organic waste.

Management. The slurry was then fed into the Newtown Creek Wastewater Treatment Facility (WWTF) during the final digestion stage in the facility's egg-shaped digesters. The anaerobic decomposition of the food waste and sewage mixture produces methane, which is captured. The DEP's Anthony Fiore explained that adding food waste at up to 20% of the capacity of the digester creates a 121% increase in gas output, with a 13% increase in methane content. While Newtown Creek had previously put its produced gas to use providing part of the power to the digester, increased methane content would allow facilities to extract this methane for other purposes. Promotional materials from National Grid depict AD as part of a new "sustainable" natural gas portfolio, in which it functions alongside fracked gas and biofuels (National Grid 2015). However, when a Greenpoint resident suggested that neighbors could get free gas from the pilot project, the representative from National Grid explained that the amount of natural gas that the Newtown Creek WWTF can provide is "but a grain of sand in the ocean" of Brooklyn's demands for natural gas.

Officially, there is no relationship between the Newtown Creek pilot and the curbside organics recycling pilot. Aside from a short-lived experiment with food waste from farmers markets,[22] DSNY has not been involved in the development of the Newtown Creek anaerobic digestion system. Nonetheless, DSNY officials and reports often identify anaerobic digestion as a possible component of organics recycling. This has created a situation in which it is difficult for officials to rhetorically separate the DSNY organics collection pilot from the AD exploration project, and concerned residents often link the two programs together (Nilsson and Warerkar 2015). It did not help that the Newtown Creek pilot was being announced at the same time that curbside organics collection was expanding to include Greenpoint.

The Greenpoint residents in attendance, having been burdened for decades with a disproportionate share of the city's waste, took issue with all aspects of the plan. They challenged both the added burden of rotting food scraps at their curbsides, and the added burden of more trucks and feedstock at the Newtown Creek WWTF. When one resident asked how organics collection and the AD pilot were related, Bridget Anderson explained that they were two entirely different waste streams—residential and commercial. This last question inspired a popcorn of questions from the audience: "Where is *our* trash going?" "Where is the digester trash coming from?"

22 This caused controversy for food scraps drop-off users who did not want their organics to be mixed with sewage sludge.

The AD future described by Waste Management, National Grid, and the DEP relies on trucking materials to the CORe facility and then to the Newtown Creek plant. But another model of AD has been available: sending food waste to WWTFs through in-sink garbage disposals. Kendall Christiansen, a consultant to the garbage disposal manufacturing firm InSinkErator and a member of the Brooklyn SWAB, has presented detailed testimony on the promise of in-sink disposal at several organic waste hearings. In-sink disposals were banned in New York City after they became popular in the 1950s, but have been allowed in new buildings since 1997. The DEP, though, has explained that garbage disposals strain the sewage system: added volume contributes to the possibility of combined sewage overflows, and sending food waste through the same processing as wastewater treatment demands more energy than any other method of disposal. DEP director Anthony Fiore has been explicit in his discussions of AD: the technology is carbon-negative only if food wastes are trucked to the facility and added to the final stages of waste water treatment. Otherwise, the process releases more carbon than it saves (Fiore 2015).

Nonetheless, in-sink disposal has gotten traction. There is a possibility that in-sink disposal of food waste would require fewer truck routes, reducing the burden on environmental justice communities. And it doesn't require nearly as much education or infrastructure redesign as curbside compost collection because it makes use of preexisting appliances and facilities. I've seen audiences *literally* hoot and holler when Christiansen explains how garbage disposals could fix the food waste problem. And in a 2016 report, the Citizens Budget Commission of New York City recommends InSinkErators as at least a partial solution, where curbside collection is not efficient enough. When I asked a sympathetic DSNY representative about the popularity of InSinkErators, she shrugged and explained that its lobby is very powerful. At the very least, garbage disposals could turn food waste into a different kind of problem: not garbage but sewage, and no longer DSNY's responsibility.

COMPOST AND THE PANDEMIC

In March 2020, New York City became the epicenter of the global COVID-19 pandemic. By mid-April, over 10,000 residents had died from COVID. The city rapidly mobilized resources to create clinics, testing facilities, food support programs, and other resources.

Based on the exodus from the city, the loss of income from tourism and entertainment, and the costs of emergency measures, city budget offices projected a $7.4 billion loss in revenue for the 2021 fiscal year.[23] While the city was able to draw on some emergency funds, Mayor de Blasio and the city council slashed $2 billion in city services (Rubinstein and Mays 2020). Among the cuts was $24.5 million in funding for DSNY's curbside organics collection and the New York City Compost Project. DSNY announced the pause in curbside organics on April 17, promising that this was a temporary measure. After the last collection, they said, residents should "use soap and water to clean your bin and place it in storage for when the program resumes" (Rubinstein and Mays 2020). There were no disruptions to garbage or other recyclables collection services.

In his daily briefing on April 17, de Blasio defended the cuts: "Things we would love to focus on in peacetime, we don't get to focus on in wartime" (Rubinstein and Mays 2020). DSNY Commissioner Garcia called the cuts "just really awful," but also explained that curbside organics "doesn't pay for itself" (Crunden 2020). Critics argued that the cuts to the composting program revealed that the de Blasio administration had never taken organics recycling seriously. Council member Antonio Reynoso, chair of the Committee on Sanitation and Solid Waste Management, said, "It just speaks to the priorities of this administration and whether or not they were committed to organics ever, if the first chance they get they want to scrap every portion of the organics program" (Muoio Dunn 2020).

The curbside organics pilots had already been thrown into question in the two years before the pandemic. Expansion of the pilots stopped in May 2018, in part due to low participation. In her testimony to the Sanitation Committee, Commissioner Garcia explained that DSNY wanted to "ensure New Yorkers are getting the very best service when curbside organics collection reaches their neighborhood" (Committee on Sanitation and Solid Waste Management 2018). Council Member Reynoso explained the decision from a budgetary perspective: "They have trucks that are moving around half-empty—that is not good in a cost-benefit analysis. We're willing to take a hit in an effort to ease the organics program ramp-up, but not so much of a hit that we're just bleeding money and tax dollars" (quoted in Muoio Dunn 2018).

23 The fiscal year ran from July 1, 2020 to June 30, 2021. For reference, a typical annual budget for NYC is around $100 billion.

While the pilots served about 40% of New Yorkers, they collected only about 5% of the city's food waste. And while organics collection was supposed to save the city money, the low participation rates, high tipping fees at industrial composting facilities, and the lack of a lucrative market for finished compost meant that organics collection was not cost-effective. The organics collection program cost $15.7 million in 2018, but the city was able to recoup only $58,000 in compost sales (Collins 2018). The Citizens Budget Commission of New York, an independent watchdog organization, recommended in September 2018 that "the curbside program is not ready for citywide expansion" because of "unreasonably high" costs per ton for collection, and a lack of local processing capacity. Figures from New York City OpenData indicate that organics diversion rate dropped slightly between 2018 and 2019 (DSNY 2023).

Nonetheless, defunding the program drew a strong public response. Community composters, microhaulers, environmental justice advocates, and compost enthusiasts came to the defense of organics recycling. The Alliance for a Greater New York, a community and labor organization, held a #SaveOurCompost Town Hall on May 5, the day after organics collection was suspended. Over 1,000 people attended the virtual event. Panelists included microhaulers, community composters, city council members, the zero waste consultant who had started the original "Save Our Compost" petition, and Manhattan Borough president Gale Brewer. Sandy Nurse, founder of BK ROT, was the first to speak, and highlighted the importance of curbside organics in her remarks: "We don't have the luxury of time to push composting at large scale, at the city scale, further down the line." Speakers highlighted the importance of both citywide and community composting.

The Save Our Compost Coalition continued to defend the curbside organics program, alongside community composting. In a hearing for legislation that would restore funding to community composting, 35 people—many from Save Our Compost—spoke in defense of curbside collection. Dr. Tok Oluwaseyi Oyewole of the NYC Environmental Justice Alliance framed the cuts to curbside collection as an environmental justice issue: "Completely eliminating all forms of organics collection means that the majority of this waste will go to the intermediary local transfer stations then to landfills and incinerators in EJ communities outside of the City." Others framed it as a problem of uneven access to composting opportunities: Emily Bachman of the NYC Community Composting Coalition said that "the proposed budget creates a landscape where composting is only available to privileged New Yorkers with the ability to pay private companies to do the work our City should do." Dior St. Hillaire of the microhauler GreenFeen OrganiX likewise explained that some people "may not physically be able to make it to drop-off locations whether it is because of disability or time constraints." And many framed organics collection as part of the city's climate goals. Justin Green of Big Reuse, a NYC Compost Project host site, explained that "mandatory curbside collection and community composting are essential and should be reinstated immediately as part of our responsibility to fight climate change" (Committee on Sanitation and Solid Waste Management 2020).

Many of the speakers demanded that curbside organics be reinstated with the same mandatory participation as curbside recycling. The Save Our Compost Coalition included mandatory organics recycling as one of its three demands: "Introduce legislation to set a timetable for a mandatory curbside organic waste collection program, to make organic waste recycling accessible to all New Yorkers in their homes and workplaces. This is necessary to meaningfully reduce the greenhouse gas emissions from waste, create local green jobs, and reduce air pollution in low-income communities and communities of color" (Save Our Compost Coalition 2020). City council speaker Corey Johnson had included mandatory organics recycling on a list of his climate priorities in March before the pandemic hit. And it was a popular policy proposal in the 2021 democratic mayoral primary: seven out of the eight front-runners said they planned to make organics recycling mandatory.

While de Blasio had courted the idea of mandatory organics, he found it politically fraught. In a 2021 press conference, he explained that "the question of mandatory to me is a challenging one in terms of what we can ask of people and what's going to work." On April 22, 2021—Earth Day—mayor Bill de Blasio announced that curbside organics collection would be reinstated the following fall, with some modifications: instead of picking up from every household in a collection zone, buildings in collection zones would sign up for organics collection. In a hearing in May 2021, DSNY commissioner Eric Grayson explained that the opt-in system would save the department money and truck routes: "Now we can really target organics collection particularly in the residential areas to the clientele and the constituents that want it and want to get directly involved so we can start diverting as much as possible from the buyer and the stakeholders who are already bought in while we continue our outreach plan" (Committee on Sanitation and Solid Waste Management jointly with the Committee on Finance 2021). The opt-in plan built on recommendations from the Citizens Budget Commission, which encouraged DSNY to implement pickups only in zones with high diversion rates (2016).

However, critics of the plan felt that it moved away from the city's diversion goals, and its responsibility to address climate change and waste inequities. At the same hearing, Council Member Reynoso said to Commissioner Grayson, "We know to get to zero waste, we need this compost program to be mandatory and to be citywide. And it seems like every time we have a conversation about compost, we are either taking a step back or taking a long time to push it forward. Eventually, the Department of Sanitation has to be committed to get to zero waste . . . I just don't feel like in the eight years I have been here that we have taken it serious" (Committee on Sanitation and Solid Waste Management jointly with the Committee on Finance 2021). Kathryn Garcia—who at this point was running for mayor—likewise railed against the plan: "It is going to turn composting into a 'luxury' that is available for New Yorkers that have the resources to organize community support and submit bureaucratic paperwork. Curbside organics should be universal—plain and simple" (Duggan 2021).

Grayson's point about efficiency should not be minimized. Running empty trucks through neighborhoods with low participation rates causes extra airborne pollutants and greenhouse gas emissions. Matching routes with resident interest, Grayson said, would allow the department to "run the appropriate amount of truck shifts for the most outcome." It's not a given that an inefficient organics collection system powered by fossil fuels will emit less carbon equivalents than a system that sends all its waste to landfills.

Furthermore, mandatory collection comes with its own set of problems. Existing inequities in access to recycling and composting opportunities, as well as in policing, could mean that people of color and low-income residents are more likely to be targeted for enforcement. And mandates are usually enforced with fines, which tend to operate as a regressive tax. Kristen Munro frames sorting recyclables as a form of unpaid household labor that overproduces resources for recycling companies, without any compensation to households (2021). While mandatory organics recycling promises higher rates of participation and diversion, it does not necessarily promise justice for waste generators.

However, the idea of voluntary participation makes sense in the first place only if organics recycling is thought of as a service for residents. The conflict between mandatory and opt-in organics ultimately comes down to a big question about municipal compost: who is organics recycling for? For the Save Our Compost Coalition, separating out organic waste helps to create a more sustainable city for everyone, and to decrease the city's impacts on the areas that deal with New York City's waste or on the earth's climate. For de Blasio's opt-in composting program, composting is a service that is meant to benefit residents—"clientele"—who want to separate out their food scraps. In the opt-in system, the individual right to throw away what you want, how you want, is more important than communities' rights to clean environments.

With the opt-in plan, we see another element of neoliberal waste management, in which individual waste generators are given control over a decision that could instead be made as a community. While self-determination and infrastructural efficiency are important, neoliberal waste management makes it easy to forget who has to live with the city's wastes.

CONCLUSION: COMPOSTING NEOLIBERALISM

The forms of organics recycling that have come to dominate in New York City are resonant with the dynamics of neoliberal waste management. Organics collection, organized by the city, has created a new waste stream that the city manages through contracts with private haulers. Individual households, imagined as consumers of waste services, can choose whether or not to participate in organics recycling. Contamination in the diverted material makes it difficult to create a waste stream pure enough for meaningful reuse. In this context, the market for organics hauling and processing favors waste management approaches to composting over agricultural approaches. Arrangements in other US cities have followed a similar pattern (Doersam and Gregory 2016; Stevens 2011).

From the beginning, the organics collection pilots have been framed as both more sustainable and more cost-effective than traditional disposal. While the promise of financial savings helped to make the pilots more popular, it also harmed the pilot when it failed to live up to this promise. This is not a new story for municipal composting. In 1972, the compost engineer Clarence Golueke explained that previous composting programs had been derailed by an unfounded confidence in the profitability of composting, and an insistence on assessing the efficiency of composting programs based on financial returns (2009). While composting could make economic sense—it saves money on disposal and creates resources out of wastes—the current economic context makes it difficult to translate this promise into actual savings.

If the curbside organics pilot is progress, it is progress toward a specific vision of organics recycling that includes opt-in participation, long distance hauling, poorly contained systems, low-quality final product, and private ownership of publicly generated resources. As actually implemented, the pilots exemplify the busy-ness that Samantha MacBride finds in other New York City recycling programs: a lot of work that feels productive but is materially insufficient to address the actual scale of waste problems.

One of my commitments in examining the curbside organics pilots is to read these efforts reparatively while avoiding the abstract utopianism of incremental thinking. How can we find promise and potential in the "small step" of voluntary residential organics collection without accepting the unlikely claim that an opt-in organics program demonstrates progress in the city's plan to meet its waste and climate goals? One way to do this is to focus on the ways people have made their own meaning out of curbside organics collection.

Separating out food scraps has the potential to change the ways that people think about their wastes. Gay Hawkins finds political significance in the activity of recycling, and the materiality of the waste stream (2006). By paying attention to the experience of sorting and stewarding waste, Hawkins offers a fine-grained analysis of recycling practices: "The dismissal of changes in practices as tokenism perpetuates the idea that politics is restricted to macroassemblages like the state or capitalism, and that real social transformation is possible only via wholesale revolutionary change" (6–7). Hawkins moves the conversation around waste away from cost-benefit analysis and diversion goals, and into the world of embodiment, habits, and feelings. If we pay attention to the stickiness of these experiences, it's easier to imagine how the habit of source separation changes people. Cultivating this habit might encourage waste generators to imagine themselves as part of a metabolism with the soil, or as part of a community of composters. It might create citizens who relate to their wastes in exciting new ways. Or it might not. But the fact that neighborhoods with organics collection increased their paper and metals, glass, and plastic diversion rates (DSNY 2015a) indicates that these habits have the potential for deeper impacts.

The curbside organics pilots also hold promise and potential as raw material that activists have been able to compost into new visions for the city's waste system. When funding for curbside organics collection was cut, community composters and environmental justice activists turned out to fight for the reinstatement of curbside organics—but this time with mandatory participation and increased support for community-based systems. In the #SaveOurCompost campaign, activists rearticulated the organics pilots not as a disposal option for the eco-conscious consumer but as part of the city's responsibility to reduce the impacts of its waste system on marginalized communities.

The Save Our Compost Coalition could be read as activists being captured by the agenda setting of industry. In MacBride's account of recycling activism in New York City, the activists demanding citywide recycling of glass, metal, plastic, and paper helped to push for a system that did not wind up meeting the stated goals in terms of waste diversion or equity. MacBride explains that industry and policymakers manipulated the conversation around recycling, and presented publics with a false choice between disposal and recycling. Neither option is particularly effective at reducing the impacts of wastes, but activists pushed for recycling because it seemed to be the more sustainable option.

With the Save Our Compost Coalition, though, activists are not just advocating for the more sustainable option, they're adding their own alternatives to the mix and changing the frameworks that are used to assess organics recycling. Activists are arguing that the organics recycling program *could* meet their needs through the incorporation of community composting and microhauling options. Curbside organics might have been a "step" in the right direction, but only because activists have redefined that direction.

CHAPTER THREE

The NYC Compost Project

LOCALLY MEANINGFUL COMPOSTING IN PARTNERSHIP WITH DSNY

In the antiseptic green classroom of the Lower East Side Ecology Center (LESEC), Ariana drew a rainbow in my notebook. I had asked Ariana, the manager of the NYC Compost Project hosted by the LESEC, how the local work of the Compost Project—food scraps collection at farmers markets, technical assistance for community composters, and training for home composters—fit in with DSNY's unfolding residential pickup project. The rainbow was her answer; she explained that the outer arc of the rainbow was municipal and commercial collection, and the inner arc was community compost and commuter compost.[1] "Everything is enveloped in the biggest component of the rainbow, but it's just in different scales."

The compost rainbow describes the Compost Project's vision for the future of organics recycling in New York City. What looks like one big project (getting New York City to compost) is actually multiple projects (household, community, and citywide compost) that all work together to create a set of changes that we see as a composting city.

Community compost in New York City emerged as a variety of informal infrastructures functioning at the edges of the city's waste system, but has been partially incorporated into a more formal structure. Community compost sites are experimental infrastructures that meet the needs of both existing and imagined communities. Projects that were built as community

1 A compost drop-off site near subway stops, open during the morning rush hour so that commuters can drop off scraps on their way to work.

spaces or as local systems have also been reframed to serve the larger project of making New York City compost.

The NYC Compost Project is the principal body working to incorporate these smaller, local systems into the compost rainbow. It is run by the Department of Sanitation, as part of its Bureau of Recycling and Sustainability. Louise Bruce, then the program manager, explained that its goal is to build a "culture of composting" in New York City through education and outreach, and by supporting compost demonstration sites.

While the NYC Compost Project serves DSNY by encouraging participation in curbside organics pickup and other forms of diversion, it is also a resource for community composters and gardeners. This makes its relationship with the informal infrastructures of community compost more complicated: part of its function is to shape community composting in service of waste diversion, but the Compost Project has also worked to understand and learn how to support community composters. The NYC Compost Project is what David Guston refers to as a *boundary organization*: it encourages the development of boundary objects (such as compost piles) and standardized packages (community compost best practices) that can fit into the goals of both waste management and urban sustainability (2001).

This chapter describes the NYC Compost Project's role in the transition to organics recycling, its relationship with the dominant approach, and its support of concrete utopian efforts to reimagine the waste system. Following a legacy of paranoid readings of recycling and other waste reduction efforts, I entered the field assuming that the Compost Project's position as a boundary organization was not politically neutral. In writing this chapter, I paid special attention to the ways in which DSNY has the upper hand in its interactions with grassroots compost, and has used existing informal systems to support the emerging centralized organics recycling regime. However, I also looked for the ways in which the project has opened up spaces for alternative systems that can nurture experimental relationships with waste and space. The relationship between DSNY and community composters, as facilitated through the NYC Compost Project, demonstrates a tension in neoliberal waste management: alternative systems can simultaneously challenge the regime and be captured by it. In the conclusion, I'll describe the dual role of community compost sites in New York City, both furthering the dominant waste management regime and articulating new meanings and designs for urban compost systems. I also suggest strategies for materially supporting these projects in order to make these utopias more concrete.

THE ORIGINS OF THE NYC COMPOST PROJECT

When DSNY began its first food scraps collection programs in 1993, it included an educational component: brochures, household posters, and a compost hotline. DSNY contracted with each of the four borough Botanical Gardens—Brooklyn, Queens, Staten Island, and the New York Botanical Garden in the Bronx—to host educational programs about composting. DSNY reasoned that if it was going to ask New Yorkers to do the dirty, smelly work of separating their food scraps from the rest of their garbage, it needed to explain why this new chore was useful for residents, for the city, and for the environment. This cultural facet of DSNY's composting effort was called the Compost Project (Lange 2001). After the curbside organics pilots were canceled in 1996 due to lack of buy-in, DSNY continued supporting the Compost Project so that it could run educational programming. Louise explains that DSNY understood the failure of the pilots to be in part due to a lack of understanding, and a perception of compost as "gross."

Louise served as my primary guide to the Compost Project. She has a seemingly limitless excitement about composting, and a knack for figuring out how to make composting happen in all the different nooks and crannies of the city's waste system. Every time I spoke with her, we found ourselves dreaming up new schemes that could help to solve this or that element of the city's waste problems.

Louise explains that, for much of its existence, the "bread and butter program" of the Compost Project was the Master Composter Certification Course. The project was a "train the trainer" course, aimed at giving people the skills to help others compost at home and in community gardens. The course was modeled after similar programs in other cities that had existed since the 1980s. In 2012, the Compost Project launched the Local Organics Recovery Program, which funded drop-off sites and processing locations through three nonprofit organizations: the Lower East Side Ecology Center, Big Reuse, and Earth Matter. DSNY provides another funding stream for greenmarket composting through the nonprofit GrowNYC, which operates most of the city's greenmarkets.

The growth of the Compost Project, starting in 2012, complemented the shifting focus on recycling and sustainability at DSNY. Through most of its life, the Compost Project was small enough to be invisible to other DSNY employees, even within the Bureau of Waste Prevention, Reuse and Recycling (BWPRR). Louise explains that in the early days of the NYC Compost Project, it was "significantly underresourced for the scale of the city they aimed to serve." Nonetheless, it was able to develop "creative, lasting programs that serve as the foundation for today's sustainability agenda." BWPRR staff credit the small size of the project for its longevity: as one of the smaller-ticket items in the DSNY budget, the Compost Project was not an obvious target for budget cuts. The status of the bureau changed during the de Blasio administration, with OneNYC's "0X30," and the appointment of Kathryn Garcia as commissioner of DSNY. Rebranded as the Bureau of Recycling and Sustainability, the former BWPRR became a much more powerful and visible office at DSNY.

"HOSTED BY": PARTNERING WITH DSNY

When I was conducting my fieldwork, between 2014 and 2016, the Compost Project employed a program manager and project coordinator who worked at the DSNY offices on Beaver Street, and a staff of 37 employees spread across seven host sites. Marguerite Manela, who became program manager after Louise in 2015, explains that the distributed structure is intended to help local knowledge influence project activities: "By design, the staff are staff of different institutions throughout the city, in different geographic locations, and the reason for that is for them to use their local understanding of their local communities to direct their work."

The NYC Compost Project is guided by a particular grammar, which I fumbled my first times in the field. The phrase "hosted by" (as in, *The NYC Compost Project hosted by Earth Matter*, the official name of the compost site on Governors Island) was designed by Louise and colleagues to "emphasiz[e] the partnership . . . in the name"—that is, Louise didn't want people to think of these sites as either NYC Compost Project or independent sites; the idea was to always describe them as partnerships. This results in gargantuan acronyms like "NYCCP hb EM" (for NYC Compost Project hosted by). When I mess it up around Compost Project employees, they correct me and walk me back through the proper name. The "hosted by" language was relatively new when I was doing my fieldwork.

When people at DSNY explain the beginnings of these partnerships, they use the language of discovery. Louise says that "what they saw happening was throughout the city, this really incredible network of very sophisticated programs: community gardens, community compost sites were being led by their master composters, and some of them were actually attracting quite a lot of funding." Marguerite explains that once DSNY recognized that these projects existed, it decided to "look to the communities to help us manage some of this waste but also to care for the spaces and to rebuild soils." Louise adds that DSNY recognized that these organizations had connections and social capital that could be useful for building compost in New York City: "Encouraging residents to make changes in their in-home materials separation systems at a meaningful scale is very difficult. To best facilitate a movement toward citywide organics diversion, BWPRR believed an effective tactic was to fund the organizations that are on the ground, nimble, that have the networks, that are going to be very close to the people we're serving. These 'trusted voices' would be more persuasive and effective than outreach officers centrally deployed to reach residents across such a large jurisdiction." In the following stories, I describe the ways in which two Compost Project host sites, Earth Matter and Salvage Mind Compost, mediated the interests of DSNY, local projects, and individuals as they became integrated into the Project.

EARTH MATTER NY, INC.

I first met Charlie Bayrer at a policy forum on organics recycling at New York University. In a sea of nonprofit workers in hip winterwear, and waste industry professionals in expensive, loose-fitting suits, he stood out in a denim shirt and jeans with wide suspenders—the only one there who looked like a farmer. Charlie is one of the central figures (along with Marisa DeDominicis, his partner and cofounder of Earth Matter) in the community farming scene of New York City. He speaks slowly and methodically, in the kind of complete sentences you rarely hear in an interview context. He told me the history of Earth Matter in Marisa's spacious Lower East Side apartment.

In 1996, Charlie, then a building superintendent in Brooklyn, began adding food scraps from his building to his landscaping compost heaps. Like many composters, he was amazed to witness the transformation of garbage into clean soil, and began to spend more of his time working on finding better places for organic waste, schlepping neighborhood leaves and grass clippings to nearby gardens, and providing technical assistance to gardeners building and maintaining compost piles. When he joined a garden in 2005, he worked with the gardeners to set a goal for composting based on the compost they needed, rather than the organic waste they had access to. Recognizing how much material this would take, Charlie worked with gardeners from Greene Acres and Brooklyn Bear's Community Gardens to find other sources. Following brief discussions with GrowNYC, they set up a partnership with the Fort Greene Farmers Market, where they could set up a table and collect food scraps to haul back to the garden to compost. Whenever they collected more food scraps than they needed, they tried to sell the farmers on extended producer responsibility for food scraps: "That Brussels sprout stalk is excess packaging, you should take it back." Farmers were not receptive to this idea.

Between 2005 and 2009, the Fort Greene Compost Project established itself as a reliable presence at the farmers market, biking materials to three (later four) local gardens for processing, in a rotation based on each garden's need and capacity. Charlie had begun working with Marisa DeDominicis and Kendall Morrison, two urban farmers and community activists who connected with him through the Brooklyn Queens Land Trust. During this time, their only contact with DSNY was taking Master Composter Course students on tours of their operations. At the time, GrowNYC had not yet established food scraps drop-off at most of its markets, and the only comparable model was operated by the LESEC at the Union Square Greenmarket in Manhattan. Brooklynites who had previously lugged their food waste across the bridge were happy to have somewhere local to take their scraps.

In 2009, gardeners started to recognize that they were making more compost than they could use, so the project reorganized and began taking its food scraps to Red Hook Community Farm, four miles away. By 2011, they were composting 3,000 pounds of food scraps at Red Hook per week. While the farm offered a reliable processing site that was far from residences, along with an established volunteer base, it also created new needs in terms of transportation: gas and occasional truck rentals. The project began to collect donations along with drop-offs, in order to offset these costs, but recognized that it would need a formal structure in order to make sense of its finances.

Marisa, Charlie, and Kendall incorporated as Earth Matter NY, Inc., in

2009 in order to participate in the City of Water Day festival on Governors Island, where they collected food scraps and tabled to talk about urban composting.[2] While they didn't manage to capture more than a cubic yard of paper plates, their tabling—which included a worm bin and their own infectious excitement—endeared them to the Governors Island Preservation and Education Corporation, which managed the island. In 2010, Earth Matter was offered space to set up an education station for composting food scraps produced on the island.

At the time, local composters were beginning to doubt the longevity of the NYC Compost Project. The Compost Project was not yet funding large demonstration sites, but it was running the Master Composter Course and providing educational materials. Charlie explains that part of their reasoning for establishing Earth Matter as a demonstration site was the possibility that the project would be lost in department reshuffling: "Perhaps there's a need for an outside organization to support the work that the Compost Project does, or in time there may even be a need to replace that if it goes down the tubes."

In 2011, council speaker Christine Quinn appropriated funding for food scraps collection at seven GrowNYC markets. Because there wasn't enough local capacity for the volumes expected, GrowNYC contracted with the private company, Action Carting, to carry food scraps to the Peninsula Compost Facility in Wilmington, Delaware. However, this was a politically unappealing arrangement because the people who put in the effort to bring their food scraps to the farmers market were watching them get trucked 90 miles away to be processed and sold by Waste Management, which acquired Peninsula two months into the pilot. In late 2011, the Earth Matter team raised their hand and offered to process some of this material at their new space on Governors Island.

Around the same time, the NYC Compost Project approached Charlie and Marisa to begin discussions about using their operation as a Compost Project demonstration and processing site. At this time Earth Matter had been operating under the city's radar for years, and Charlie was somewhat nervous to be approached by DSNY: "Attracting the attention of officialdom is not always a good thing." Debbie Sheintoch, then manager of the Compost Project, invited Charlie and Marisa to take part in the Local Organics Recovery Program, a new partnership between DSNY, GrowNYC, and

2 This is an annual event that celebrates and raises awareness of NYC's waterways.

local community organizations. The Earth Matter team designed a project that focused on diverting food scraps, compost education, and a zero-waste island, and also found local uses for finished compost in both farming and landscaping applications. Charlie credits the local uses for compost with making the project especially appealing to DSNY: "You're not gonna produce compost that can be sold by itself, with any great return, but if you add the value of food production to it, you've increased the worth of your effort and you've got a more politically saleable outcome." Earth Matter became a host site for the Compost Project, and the department created paid positions for Charlie and Marisa (and eventually others) to build a demonstration site on Governors Island. Their facility is now open to the public on weekends in the summer, and for weekly volunteer days the rest of the year.

SALVAGE MIND COMPOST

"P." described themself as "too big and too brash" to get into the local composting world of New York City.[3] They *are* big, with the kind of boxy body that doesn't quite fit comfortably into the cramped Lower East Side apartment where I first met them. There's a kindness and excitement about them that gives me the impression of unbridled energy that they're struggling to contain. From their stories about the beginning of What a Waste! Compost, it sounded like the early days of the project were their effort to find an outlet for their superhuman energy supply. They told me about spending nights turning thousands of pounds of compost by hand while listening to classical music. I wondered how much of the infrastructures for handling excess materials in New York City were built by people trying to do something with their own excess energy.[4]

In 200-, P. joined the Growing Strong Community Garden in Bedford Park in the Bronx. D., another garden member, was working to increase composting at the garden, and she roped P. (fresh out of the Master Composter Certification Course) into building bins for the garden and helping to rally gardeners. Like Charlie, they were interested in creating a compost system for the garden that wouldn't just handle the waste generated on-site,

3 This name and all organizations within this subsection are pseudonyms. I tried to match the culture of compost/plant puns that community composters enjoy. Like composting, punning is a practice of finding new meanings in existing material.

4 People with too much energy aren't unusual in local compost. Louise gives this impression too, with her big eyes and excited talk. Domingo from BK ROT regularly jokes about his Energizer Bunny heart while he tosses his extra-large shovel around like it was a grapefruit spoon. It's not that everyone in local compost is larger than life, but there's a symmetry between the informal economies of excess material and excess labor.

but could also generate compost enough to be useful for the whole half-acre garden. Urged on by funders such as the Citizens Committee, they tried to design a compost system that could go beyond the garden to include the surrounding community.

P. and D. developed a system modeled after the Fort Greene Compost Project, in which volunteers picked up food scraps from two farmers markets and delivered them to a central site on bicycle. The site was originally at Growing Strong, but P. soon found a larger site on underused Parks property and called it What a Waste! Compost. By the following summer, they were picking up 2,000 pounds of food scraps per week. D. stepped back, but H., another master composter, began working with P. and helping to find grants.

In 201-, they received a large grant that required a match from a nonprofit sponsor. What a Waste! partnered with Salvage Mind, an organization that had been offering nominal support to What a Waste! in terms of materials and space. The grant, over $100,000, was enough to cover materials and two entry-level salaries, and H. came on to run the organization with P. They set up more collection sites at farmers markets, and spent summer nights turning compost (the nocturnal schedule was meant to cut down on the smell). They set up organics collection at bike races to promote recycling and waste diversion to audiences that weren't traditionally considered "green."

After about a year of funding, P. and H. were approached by Debbie Sheintoch, who offered to fund their work through the Compost Project. Because What a Waste! wasn't an independent organization, the funding still moved through Salvage Mind, and the project was incorporated into Salvage Mind as Salvage Mind Compost.

One of P.'s goals with this project was to create a kind of "systemic change" that they didn't see any of the other compost projects in New York City working on. They thought it was a problem that the city was moving toward a "client-oriented" curbside collection system, and wanted to create a system that helped people to interact with their waste more: "I thought it was important to take compost to a central place and to take the mass of material as a reference point. When it's on the curb, it's easy to ignore it. Once you see it in the aggregate, it makes a different impression. Also, when you bring people together, it makes them talk about what they're doing. Giving it a form where it allows people to give a shit is maybe not always going to be acceptable to the status quo because it makes their jobs harder."

All of the Compost Project sites rely on a system where people come to central places to drop off their food scraps, and in most of them you can see and smell the waste. It leaves an impression, but this impression is nearly always one of achievement and order. The GrowNYC food scraps drop-off at Grand Army Plaza has two straight lines of bins and toters that extend up the hill into the park. The smell is pleasant, fruity, occasionally floral. P. wanted their compost system to suggest something messier, something a little harder to deal with. They were frustrated with the poetics of neatness and control that were embedded in most other Compost Project infrastructure.

Before the COVID-related funding cuts of 2020, I struggled to find places where the interests of a host site conflict dramatically with the interests of DSNY. While the host sites are all very different in terms of their approaches to labor, technology, publics, education, and space, they all maintain relationships in which the goals of the two organizations seem to overlap very neatly. Aside from some personal disputes, the only complaint I've been able to elicit on the part of NYC Compost Project people working at host sites is that bureaucratic red tape makes it difficult and time-consuming to requisition new materials. Of course, this all changed with COVID-related funding cuts, which I discuss later in this chapter.

WHOSE LABOR POWERS COMMUNITY COMPOST?

By December 2015, I was used to the ferry ride to Governors Island—I knew not to pay any attention to the e-mails warning me to arrive a half hour early, and I knew I needed to find coffee beforehand because there'd be nothing on the island. Andrea, who coordinated the volunteer days, recognized me as I stood with my bike in the waiting area for cars. She called to me and said hello but stayed inside the passenger waiting area to keep an eye out for less seasoned, more confused volunteers.

It was a windrow build day at Earth Matter, then a monthly event where GrowNYC drops off six tons of food scraps, and a ragtag crew of volunteers shows up to dump them out into a new windrow. I had been to some of the Friday volunteer days, where I fed chickens, shoveled scraps, harvested vegetables, and moved compostable cutlery around. I was curious about the windrows, and how they got there.

The windrows at Earth Matter are about seven feet high and 30 feet long, with steep sloping sides. They are covered with woodchips and give an impression somewhere between architecture and life-form. They have the mysterious presence of an ancient tomb: quiet, looming, not exactly alive, but hard to ignore. For farmers and composters, a big windrow like this might evoke what David Nye calls the "technological sublime"—that feeling of transcendence that viewers get when they stand in the presence of human-made objects that seem to challenge the laws of nature (1994).[5]

It was cold on the island, and we had only a handful of people. Maybe five or six. One was a young woman I'd met at a mixer for waste professionals the night before. Over crudités, she had slyly and proudly told me how she buries her meat scraps in Prospect Park because she can't compost them at home. Another was Travis, a young man from Rockaway who spoke in Marxian terms about his former job in the Parks Department. We also had the Earth Matter staff (Marisa, Charlie, Andrea, and the site manager, Syed) as well as two of the three apprentices (Infinitea and Pat; Frederick was absent). Two men from GrowNYC showed up to drop off the scraps, including the driver, Max.

The GrowNYC men didn't want to hang around; they unloaded the bins into three neat rows, each stacked three high, that must have been about 25 feet long. Marisa had us find partners to lug the bins over to a bed of wood chips that Charlie and Syed had set up, and dump them out. I worked with the renegade meat composter. We didn't talk much, but we commented on the things we saw in the scraps: flowers, fennel bulbs, whole loaves of bread. Our mouths watered for the discarded food. It was much fresher than the stuff I was used to seeing at other sites; it still smelled edible. While we dumped the bins, Charlie and Syed operated a toter tipper, a yellow mechanical device that lifted large toters of food waste (like a plastic curbside garbage can) about six feet into the air and inverted them, spilling their contents into the pile. Collected from Grand Army Plaza (a farmers market that serves the bourgeois neighborhood of Park Slope) in winter, many toters were loaded with beets, and the juice dripping out of them looked like blood running down their yellow lips.

The bins were heavy, and my back was sore, but the labor didn't last long,

5 I hear people marvel at compost piles. Charlie tells me that "once you discover the magic of the transformation of organic waste into usable soil amendment, it sort of grabs on to you." The transformation of waste into soil is alchemical, and for many composters, that miracle is what keeps them coming back. I don't always feel this enchantment. But when I'm in the presence of big windrows, and can feel their heat and understand their size, I start to get that magical feeling.

FIGURE 1. Charlie and volunteers watch as the toter tipper dumps out a toter on a new windrow, 2015. Photo by author.

and soon Marisa instructed us to chop and spread. We didn't need to julienne the whole expanse of leftovers, she just asked us to try to make an even layer, to make sure big things like winter squash and pineapple tops were broken up a little bit. My partner and I picked up flowers from among the scraps, held on to them for a moment, then dropped them. Nobody could tell when we had chopped enough. We idly whacked pumpkin halves with choppers waiting for someone to tell us to stop.[6]

Andrea gathered us back into a big loose circle surrounding the thin layer of food scraps on the ground, and moved us through the formalized process of thinking up a name for the windrow. All the other windrows had funny, often dated, names written on on a little plaque with the date: "Cecil the Lion," "Pumpkin Spice," "Schedule Change." It was Earth Matter's way of recognizing volunteer labor: the Master Composter Manual (MCM) recommends making volunteers feel like they're an important part of the project. Today there weren't a lot of ideas; someone suggested "Jimbo Jimbo," but it didn't receive much support. I suggested "Fearless Falcon," referencing a story that Charlie had told in the morning's go-around, and everyone seemed enthusiastic about that. Andrea shouted out the two names and asked for a show of hands, Fearless Falcon won and the windrow was named.

Andrea split the group up to do other tasks, like helping to build the compost toilet or the rabbit hutch. Meredith, who had shown up to work with the animals, led me and Travis through the process of chicken care. She was dressed in several layers of flannel, and spoke energetically. I asked about her involvement in the project and she said that this was a volunteer thing for her; she loved animals and organic farming, and took a day every week to come care for the chickens. Travis and I shoveled up all the chicken droppings from the hoop house and put them on the pile of old food scraps that Syed was taking over to the freshly built windrow. We laid down new wood shavings for bedding, and then took the chickens' roosts out of the hoop house to clean them off with vinegar water. They didn't seem very dirty but we couldn't get them very clean. Travis and I didn't know if our scrubbing accomplished anything.

Once the work was done, Marisa led us on a tour of the different compost systems that had been built by the apprentices. She asked us questions about them as she led the tour, watching to see if we'd learned anything during our time there. We all walked together back to the ferry, talking about the things we'd seen that day.

Charlie described the volunteer days as "the easiest way to plug in with us and find out more about composting." Earth Matter has an online signup form for volunteer days. Until the beginning of the pandemic, the website

6 Choppers are sold as "edgers" for landscaping purposes—basically a flat metal semicircle with a sharpened circumference on the end of a shovel-long handle. For years I thought they were called "choppers" because the tool and the term are so ubiquitous at community compost sites.

listed opportunities ranging from the current week to the end of the year, along with how many slots were left out of twelve. Volunteers were asked to give their name and contact information, how they learned about Earth Matter, and whether they were in a Master Composter Course. Volunteer days were paused with the beginning of the COVID-19 pandemic, and visitors to the website were instead encouraged to sign a petition to help restore funding for community composting.

Once you signed up, the volunteer coordinator would send one or two automated e-mails about your upcoming shift, including directions to the site and instructions about what to bring. It was still easy to forget that you signed up, though, and generally a few sign-ups fail to make the 8:30 ferry, or don't show up at all. I've never heard of any consequences for not showing up.

Earth Matter makes it very easy to work at its site because volunteer labor (aka experiential learning) is an important component of its educational project. The facilitators go out of their way to make us feel like a team, and tasks are split up in a way that encourages cooperation. The bins that seem universal to Compost Project sites are awkward and heavy, and best moved by two people. The windrow-building labor is split into the lower-skilled task of hand-dumping bins of scraps and the higher-skilled task of using heavy machinery to cover the scraps with mixed scraps and browns. This way, our untrained bodies could get involved. The volunteer labor at Earth Matter is designed to feel good, to be educational, and to not stress out anyone too much. While I've heard stories of shifts that were too cold (the site doesn't offer any protection from the elements), a shift is generally too short to feel anything but fun.

Out of the other big seven Compost Project sites, only Added Value Compost at Red Hook Farm has such open and regular volunteer days. Big Reuse used to rely on volunteer labor for building windrows but now manages everything with paid labor and machines. GrowNYC has a volunteer-based system for collecting food scraps at farmers markets, and brings in volunteers to help people dump scraps and answer questions about composting. And most of the smaller sites invite people to come help out. Some even have systems for managing these sign-ups.

WHEN I SHOW up at the Big Reuse compost site under the Queensboro Bridge to talk with Leah, I am quickly roped into working. This is not an unusual experience at compost sites, and I'm often handed a pitchfork or a chopper in response to my "Hello." Leah, the project manager (white, with bright eyes and a thoughtful Midwestern attitude), and Charles, the operations manager (tall, Black, focused), have been dumping bins of food scraps from a pallet into the shovel of the skid-steer, then Charles has been dumping the shovel into a Jaylor to mix with browns.[7] About six bins to a shovel, two shovels to a Jaylor, and then a shovelful of browns from the leaf pile. When I add my body to the mix, Leah and I slowly establish a rhythm of moving the bins, where she removes a lid from the next one while I knock the scraps off the last one and Charles waits in the skid-steer to transport the load. Then, while he runs the Jaylor to mix up the browns and greens, Leah and I dry-clean the bins with wood shavings, pouring in a handful of shavings and rubbing them against all sides of the bin to sop up excess moisture. I screw things up: I put the bins in the wrong places, and at one point I knock a bin too hard and crack the rigid plastic. As I get tired from moving the bins, I start to get clumsy, dropping apples and oranges off the shovel and letting them roll away. Charles doesn't seem annoyed, though, and he calmly explains how to do it right.

The Big Reuse site is the most technologically advanced I've seen in the city. In addition to the Jaylor, skid-steer and toter tipper, it has a computerized aeration system that keeps track of the temperature and oxygen content of the pile and supplies more air when microbial activity dies down. The windrows are covered with enormous, thick, shimmering Gore-Tex cloth, which keeps the smells in and the rain off. Charles can maintain the site almost single-handedly, according to Leah, processing 10 tons of food scraps per week with half-time support from various staff (DSNY 2014). When the toter tipper wasn't working for two months this fall, this meant that Charles had to tip 250-pound toters by hand into the lowered bucket of the skid-steer. Leah said that this was "something that a twenty-six-year-old strong man can do but doesn't make sense for many other people, and doesn't make sense in the long run." Heaving impossible weights is a normal cruelty/machismo inflicted on "san men," explains Robin Nagle, and DSNY has earned a reputation as "New York's Strongest" (2013). In Nagle's account of life in DSNY, sanitation workers take great pride in their ability to lift enormous amounts of waste, which is a

7 The Jaylor is an animal feed mixer that has been repurposed for mixing browns and greens together. Charlie Bayrer figured out this application, and Big Reuse picked it up. The machine is essentially a giant blender, the kind that's specifically designed for slushy drinks with a spout at the bottom; a six-foot-tall plastic bucket with two rotating blades inside and a door on the side. Food scraps and woodchips go in at the top, are mixed together, and come out at the bottom.

necessary part of the job. Partly because of this, sanitation workers have one of the highest rates of on-the-job injuries (Newman 2016; NIOSH 2012).

On other days, Charles tends the piles with the skid-steer, lifting a forkful and dropping it to mix and aerate the pile, or moving the first windrow onto the second, nonaerated windrow, or moving the second one into a pile to cure. While Charles is proficient with the current setup, Leah explains how the next site will be even easier to operate, with underground vents for aeration so that Charles doesn't have to work around the PVC tube.

Even though the Earth Matter site and the Big Reuse site have similar systems and tools (static aerated windrows, skid-steer, Jaylor, and toter tipper), they differ markedly in how people use them, and what kinds of labor go into them. For Earth Matter, divvying the labor into tasks done by hand and asking a group of people to work on them at the same time allows labor to be nontaxing, educational, and even fun. "Plugging in" is available to many, which makes the experience of an urban compost system available to many.

Nina Eliasoph argues that volunteering can be counterproductive to social change—a Band-Aid that makes volunteers feel like a problem is being addressed without fighting underlying inequalities. However, she also suggests that it can also lead volunteers to question the world as it is, and "charity" projects can support conscious efforts at social change (2013). At Earth Matter, as with many other volunteer organizations, volunteering is an important part of the project both as a means of furthering the organization's goals and as a means of exposing more people to the magic of composting, and the idea of local resource cycles.

Labor at Big Reuse, by contrast, is hidden, focused, and carried out mostly by Charles, occasionally with assistance from Leah, and rarely with assistance from nebbishy ethnographers like myself. The design of the system maximizes the efficiency of the labor, to the point where Big Reuse doesn't need the kind of massive volunteer base that Earth Matter relies on. The heavy machinery also makes it more dangerous, so that only a few tasks, such as bagging compost and tree care, can be taken on by volunteers. The bulk of the volunteering that goes into this system is the labor of separating out food waste, living with it in your freezer for a week, and bringing it to a drop-off site.[8] However, this kind of labor is important in Big Reuse employees' understanding of community.

8 This is not insignificant, and Kristin Munro points out how recycling systems rely on uncompensated household labor to create raw material for appropriation by recycling companies (2021). However, this labor is significantly less demanding and time-consuming than waste processing.

FIGURE 2. Leah pilots a driving Jaylor full of mixed browns and greens at the Big Reuse facility under the Queensboro Bridge, 2016. The driving Jaylor is one of many technologies that makes the Big Reuse site easy to run with minimal labor. Photo by author.

At a meeting of the Community Composting Council, a long-awaited announcement from the Business Integrity Commission (BIC) revealed new regulations that would allow certain community compost sites to charge for pickup from commercial businesses without obtaining a permit. The eligible sites are those that either have nonprofit status or operate in partnership with the city. BIC is a regulatory and law enforcement agency tasked with monitoring private waste management (as well as food wholesalers and gambling institutions) in New York City. It was created in 2002 to discourage the kind of mob-fueled corruption that was rampant in the hauling industry until the 1990s. Previously, BIC had required an expensive permit for anyone hauling commercial waste, which made it difficult for small businesses to legally haul organic waste. Louise explains that some compost haulers had worked around this by framing organic wastes as a "resource."

The new regulations were met with cheers from the assembled mass of master composters, compost site coordinators, and nonprofit employees. But two people who ran small, for-profit hauling operations, Laura Rosenshine of Common Ground Compost and Vandra Thorburn of Vokashi, were upset about the flat ineligibility of their businesses. They posed question after question to the BIC representatives, hoping to find some way to be included in the pilot.

Later, when I spoke with Ana de Luco of Sure We Can and Sandy Nurse of BK ROT, two activists in charge of nonprofit organizations that use waste as a tool for creating jobs, they were relieved at the news. Unfortunately, BIC's rate cap of $12.38 per 100 pounds of food scraps collected was way below what either organization had been charging. BK ROT was charging for each pickup, rather than by weight, in order to more closely tie the cost of hauling to the labor of the haulers. Most small, local compost haulers wouldn't be able to function within the parameters of the pilot.

There are other barriers that make it difficult to make money off of small compost operations: Green Thumb, the Parks Department program that runs the majority of the gardens that house community compost sites, does not allow people to sell things made on Parks land, including finished compost. The Master Composter Manual provides guidelines for attracting and motivating volunteer labor at compost sites but doesn't mention the possibility of charging for collection or paying workers in anything but bottled water or finished compost. And the systems for funding local compost sites, like the Citizens Committee and the Manhattan SWAB Grants, refuse to fund for-profit compost collection services. Until recently, the Compost Project did not include fee-based compost services on its list of composting opportunities.

DSNY and BIC employees have explained that these policies are meant to keep "dirty money" out of compost, and lessen the risk that haulers could use the "community composting" designation to cover for unscrupulous business practices.[9] But they also constrain noncapitalist economic arrangements that might make it possible to compensate the labor of composting. Even though some organizers have found pathways that can create jobs out of small-scale compost hauling and processing—such as nonprofit status, operating under the radar, or incorporation into the Compost Project—the regulatory, cultural, and funding landscape make it much easier to run on volunteer labor. Relying on volunteer labor serves the purpose of "plugging in" to these systems, but it also limits the forms of participation available. One of BK ROT's primary political endeavors is mobilizing an alternative economy of composting that fairly compensates waste labor, which I discuss in chapter 4.

Later in the Community Composting Council meeting, Nando from the Brotherhood Sister Sol in Harlem suggested that monitoring the hours of volunteer work that take place on these sites can be helpful in the creation of jobs, because it can help us to understand what sorts of labor are needed and what kinds of jobs they can create. The idea seemed radical to me: characterizing the work that actually goes into these informal projects in order to figure out what sort of economy it could nurture, and what kind of waste management is being done for free.

TRAINING COMPOST EXPERTS

The NYC Master Composter Certification Course is now the gold standard for master composter courses. It builds on a tradition of master composter training courses that have been designed to teach city dwellers how to manage their waste. Howard Stenn taught the first master composter course

9 Louise suggests that these exclusions might not have been intentional—just a blind spot for regulators who were accustomed to large-scale centralized waste management.

in Seattle, starting in 1984 (Tonge 1987). Stenn's course combined his own expertise as a permaculturist with the city of Seattle's interest in reducing its waste stream. He applied for funding through the city to teach courses to 50 students who could become "master composters" on call for households and institutions interested in building or improving their own compost systems. As with master gardener programs, continued volunteering was necessary to an identity as a master composter. This model was picked up by other cities, including Denver, Colorado, and several cities in Texas.

The same kind of system wouldn't work as well in New York City, where space constraints mean that at-home composting is difficult, labor-intensive, and potentially unpleasant. The 1999 backyard composting pilot undertaken by DSNY had demonstrated that educational programs and material support for at-home composting couldn't make a significant dent in organics diversion, even in neighborhoods where most households had yards.

Because household composting wouldn't work, the NYC Master Composter Course focused on creating neighborhood systems. Master composter students were required to design and implement a project, totaling 30 hours of labor, that would advance small-scale composting in New York City. Unlike previous courses that required ongoing commitment, the New York City course did not demand that master composters continue to serve their communities after training. Many master composters do, though: they continue to advocate, to build and maintain compost systems, and to plug in to larger networks of composters and gardeners. On the application form for the course, applicants are asked to explain why they want to become master composters, what they will do with their training, and what sort of community affiliations they have. In order to qualify for the course, master composters are expected demonstrate that they'll have the resources, connections, and vision to build their own systems.[10]

You don't have to ask if someone's a master composter; they'll let you know. Master composters announce it when they introduce themselves, with the same honest but strategic tone I use when I put the honorific "Dr." before my name at city council meetings. Like any form of expertise, it means different things in different contexts (Collins and Evans 2002). At a public

10 These short essays aren't the only way people get into the Master Composter Course. Frederick, an intern at Earth Matter, explains that he spoke with the administrator of the Brooklyn Botanic Garden MCCC before applying, and suggested that I do the same.

forum, someone might announce herself as a master composter to demonstrate her commitment to her local environment and community. At a co-op orientation, the same credentials might be aimed at sparking up a conversation about gardens, soils, or food waste. Other times, the credential signals belonging in a community of thought and practice, such as when people announce which site certified them, in what year. One master composter proudly said "Brooklyn Botanic Garden, 2003," when he introduced himself at an NYC Community Composting Council meeting, and his announcement was met with awe: he was a member of the first class.

So don't bother to ask—I'm not a master composter. This comes with shame in some circles. When Renée from BK ROT couldn't figure out whether finished compost needed to breathe, she quietly admitted that she'd never gone through the master composter program, and joked that she might be unqualified to run a compost project. But she waved this worry away and said that she knows enough about composting just from doing it. Nonetheless, she still asked Sandy to bring her MCM with her to try to figure that out.

Some community composters didn't go through the Master Composter Certification Course (MCCC), and tell me that they were able to compost without formal expertise. "I just knew that I was putting some of the right things together and things were happening, and I was getting this wonderful stuff," explains Charlie of Earth Matter. There's something straightforward about composting—even if you don't bother to chop, aerate, or turn organic materials in a pile, they'll eventually break down into humus. The tricks of a master composter mostly involve knowing how to streamline the process, how to make it go faster, and how to make it more palatable to neighbors.

But speed and aesthetics are important for the Compost Project. Making compost quickly without attracting rodents is vital for making compost appealing to potentially squeamish publics. Rat prevention is a central value for master composters, who have a duty to present a tidy image for urban composting. It's hard enough to make rotting food respectable; smells and rats could ruin the image of compost.

The New York City Master Composter Certification Course has become a model for compost education projects elsewhere (Platt, McSweeny and Davis 2014). The Master Composter Manual (MCM) is now available online in PDF form, and is used in other master composter courses, as a definitive guide to home and community composting. The previous compost bible, *Let It Rot*, suggests a more passive attitude toward compost: "Let common sense and the organic materials that are most available to you be your number one and number two composting guides" (Campbell 1998, 7). The title of "master composter" implies that people can "master" decomposition, rather than "let" the natural process of rot take place.

In other fields of thought and practice, the codification of alternative knowledge has created forms of counterexpertise that structure and delimit knowledge. For instance, David Hess describes how the institutionalization of complementary and alternative medicine has both provided inroads for patient groups to claim their counterexpertise and change medical knowledge and practices, while simultaneously defusing the more radical agendas of health social movements (Hess 2004). The MCM helps to define community composting as a system of knowledge, but articulates a specific definition of what community composting should be.

The manual addresses industrial composting, which "functions almost exactly like smaller scale systems—the main difference being sheer size" (4–3) and can "greatly reduce the amount of waste being sent to landfills" (4–3). It further highlights the parallels between industrial, community, and home composting by illustrating how all three share the same basic structural elements: intake, preparation, composting, curing, sifting, and distribution. While it acknowledges that the employees of industrial sites are paid, rather than volunteers, it does not mention the differences between the funding structures of industrial compost and its smaller-scale counterparts. As in the compost rainbow, the parallel between industrial and local compost collapses the different origin stories of community and industrial composting into one unified approach.

While the manual describes a variety of different systems for aerobic composting, it frames anaerobic decomposition as universally undesirable. It makes no mention of anaerobically digesting organic material into methane, and it doesn't discuss the bokashi method of anaerobically fermenting food scraps before composting, which has found some popularity with urban composters.[11] While the NYC Compost Project has interacted positively with both anaerobic composting (by sending food scraps to the experimental AD project at Newtown Creek) and bokashi (by helping to

11 The bokashi method, developed in Japan in the 1980s, pretreats food scraps with an inoculant and pickles them in anaerobic conditions for a month or longer before composting them normally. The inoculant is meant to preserve the food scraps so that they decompose more productively, without too much odor, while breaking down tough materials that might take longer to compost. Vandra Thorburn of Vokashi Compost is an outspoken advocate of the technique; she explains that it's the only way to compost in an urban setting because it cuts down on the smell and allows you to keep food scraps in a household or restaurant for longer. After handling several hundred pounds of bokashi-fermented food scraps on a sweaty summer day, I'm not sure I believe her on the smell. Nonetheless, many urban composters, including BK ROT, have experimented with bokashi.

advertise Shig Matsukawa's bokashi classes), the MCM does not directly interact with these possibilities.

The MCM is not the same as the Master Composter Course. Even though the courses rely heavily on the manual, each one is conducted by Compost Project staff, who have significant freedom over their curriculum. P. explains that when they took the course, they felt that their instructors at Brooklyn Botanic Garden "taught me things they weren't supposed to teach in the Master Composter Course," such as systemic models of waste problems. However, the ability of the manual to move beyond the NYC Master Composter Course and represent a replicable form of knowledge makes it an important text for coordinating training programs elsewhere.

There are different ways to be a compost expert.

At Earth Matter, Marisa and Charlie are running an "adjunct program" to the MCCC, a compost apprenticeship program. "We saw the need and the interest for something to sort of actualize all of that knowledge that people were getting," explains Charlie. While the MCCC takes field trips and does activities, it tends to be a "classroom" program, and the Earth Matter team wanted to develop a more hands-on educational approach. The program is aimed at training participants to operate midsize composting operations such as at Earth Matter.

Whenever I visited Earth Matter in 2015, the apprentices were Pat, Infinitea, and Frederick. I had met Pat and Frederick before the program started, when they were both coming to Earth Matter to volunteer informally. Both of them were retired, and enjoyed being outside and working with plants and soil. Infinitea is younger, and joined the program as an extension of her own food activism in the Bronx. All three are Black.

The apprentices, along with volunteers and youth interns, built Earth Matter's "Compost Promenade." Along one side of the site, in front of a row of disused townhouses, sits a series of evenly spaced compost machines: tumblers, earth machines, and so on are set up with little informational plaques. They demonstrate different technologies that can be put in place for different contexts, with different resources.

When Marisa and the apprentices walk the volunteers through the compost promenade at the end of a workday, she'll stop to ask us questions. Sometimes she asks the crowd at large, and I blurt out an answer. Sometimes she directs her questions to the apprentices. "Infinitea, what do we do if there are

FIGURE 3. The Compost Promenade at Earth Matter, 2015. Signage has been developed by apprentices and volunteers, and encourages visitors to figure out their ideal compost system. Photo by author.

flies in the pile?" Marisa has high standards for her apprentices' answers: she doesn't want something sufficient; she wants the right answer. When Infinitea suggested adding browns to the fly-infested pile, Marisa said no, you add dry material. This is a more precise answer, even though browns—carbon-rich material, like leaves or newspaper—are almost always dry.

The compost quiz—and its prescribed answers—challenged my expectation of a fleshier, messier understanding of compost systems. Compost heaps, I had assumed, required what Evelyn Fox Keller calls a "feeling for the organism"—an understanding rooted in relationship and connection, rather than control and distance (1983). I had imagined that that composting belied the divisions between systems, users, and surroundings

(Alaimo 2010; Barad 2007). I think of Sandy wrestling with chicken wire to keep rats out of the compost bins at BK ROT. I think of Nancy at Compost for Brooklyn trying to find new ways to line up the lids on compost tumblers to keep volunteers' hijabs from getting caught on them. I think of Charles ruining his back by dumping food scraps into the Jaylor himself while the Big Reuse toter tipper is out of order. Compost expertise, as I had experienced it so far, was about feelings and kluge-jobs and embodied relationships with waste.

But Earth Matter's training and the Master Composter Course are just complementary to the more embodied expertise of community composters. Many of the big names in community composting, like Charlie Bayrer and Renée Peperone, got their start without formal training. And while the master composter training course claims jurisdiction over what happens in a compost bin, community composters regularly claim their own counterexpertise about what it takes to facilitate a community compost system. Nando's understanding of the labor involved in community composting, for instance, has currency among community composters. And when the city tried to shut down community composting at the beginning of the COVID-19 pandemic, citing concerns about contact, many community composters challenged official narratives about person-to-person contact in order to secure space to keep operating. After all, the city's support for community composting rests on the recognition that community composters know something about compost and their neighborhoods that DSNY does not.

Alternative accounts of expertise highlight the expertise that comes with lived experience (Ottinger 2013; Corburn 2005; S. Epstein 1996) or reflexivity about a given situation (Wynne 1992). They might even challenge the idea of expertise altogether: Ellen Foster and Yana Boeva describe a more distributed expertise in repair collectives and feminist hacker spaces, where expertise is understood not as a hierarchy but as something collectively held among people working on a shared project (2018). Foster explains that DIY spaces will try to create opportunities for nonexperts to experiment with technology and to cultivate the comfort associated with expertise (2019). Part of the work of community composting is teaching volunteers to be comfortable with compost, training laypeople into composters.

Community composting remains a process of making do with what you have, and assembling incomplete tools into imperfect systems. In practice, the existence of compost training programs doesn't mean that untrained composters won't have a say in compost systems. But it has mobilized structuring assumptions about how compost projects should run, and what a good compost system looks (and smells) like.

FORMALIZING COMMUNITY COMPOSTING

"Please, meet each other, learn about each other's projects, network, cross-pollinate!" This is Bridget Anderson closing out a meeting of the NYC Community Composting Council. Anderson was the deputy commissioner of recycling and sustainability at DSNY from 2014 until 2022, and reported directly to Commissioner Garcia. While Compost Project staff always seem like they'd be a little happier in a community garden, Anderson always seems very comfortable in office attire.

The meeting in February 2016 was packed, overfilling the dingy conference room at the DSNY offices on Beaver Street in the financial district. But the council doesn't typically have a strong presence. The Community Composting Council had been called together five times since November 2013 to discuss the goals of community compost in New York City and to provide feedback to the Compost Project on the needs and interests of community composters. Between meetings, the council exists as a Google Group, where a few particularly active members post job listings, events, volunteer opportunities, and educational materials. When a composter raised his hand to ask Bridget Anderson what form the Community Composting Council takes between meetings, the woman sitting next to me quickly blurted out "It doesn't" and the people around her snickered. This is not entirely true; the Google group has been active for as long as the council has existed, and became an important forum for debate and organizing during the #SaveOurCompost campaign.

The Community Composting Council is one of a handful of strategies that the Compost Project has undertaken in order to understand the variety of community compost sites in New York City, and to strengthen the networks that already exist. Having recognized the usefulness of community compost sites—their wide distribution, their embeddedness in their communities, their potential as educational opportunities—the Compost Project is working to take these existing distributed local systems and partially integrate them into a decentralized community composting infrastructure.[12] The Compost Project is working to "create this bigger structure within which people can understand how to work together, how to grow, and how to be innovative," explains Louise. NYC Compost Project leaders wanted to use this structure to help them advocate for community compost as "a critical part of our organics infrastructure."

To do this, the Compost Project has been studying what is taking place on the ground among community composters, through efforts such as the Community Composting Council, the 2014 Community Composting Report, and ongoing data collection from community compost sites. The Compost Project has also undertaken efforts to support community compost by hosting networking events, running the Master Composter Course, providing promotional materials, and offering an online database of community compost sites. The Compost Project has tried to keep an open mind in its efforts to understand the field and strengthen networks. Officials have embraced an expansive and shifting definition of "community compost" so that a variety of different projects can be integrated into the network. These vary in size, purpose, structure, and leadership. Like any effort at infrastructuring, though, building this system involves making important calls about what is included and what is left out (Star 1999).

I have already described how DSNY began supporting some of the extant community compost infrastructures by creating Compost Project host sites at Earth Matter, Big Reuse, and the Lower East Side Ecology Center. These sites have played an important role in diverting waste, creating soil, and increasing the visibility of compost systems in New York City, but an even more broadly distributed infrastructure of small local compost projects has been working across the city to meet a variety of different needs.

The 2014 Community Composting Report was compiled largely by Louise Bruce, with major support from Melissa Cipollone. It represents part of the Compost Project's efforts to understand community compost and what sorts

12 When the Compost Project turned to community compost sites in 2009, Louise explains that they found a wealth of locally designed systems. Some of this can be attributed to master composters. The MCCC requires students to take on compost projects of their own, many of which involve the planning and implementation of community compost sites. But other sites were created as a matter of course within community gardens (either to manage garden or household waste or as a source of soil amendments for depleted NYC soils), at educational institutions as part of a classroom activity, or in residences to manage waste collectively.

of impacts it has on waste systems in New York City. The report is the result of surveys, interviews, and site visits conducted over the course of two years.

The report offers a picture of a widely distributed network: in 2014, 225 community compost sites in New York City were affiliated with the Compost Project, which means that they are listed as "demonstration sites" on the project website, and that they have access to some of the project's promotional materials. Aside from the Compost Project host sites, none of these sites receive direct funding from DSNY, though many receive grants from the Citizens Committee and other organizations. These sites range in size from 10 square feet (at Clinton DeWitt High School) to 20,000 square feet (at Red Hook Community Farm). Nearly half are located in community gardens; others are in schools, parks, urban farms, and apartment buildings. The majority (83%) have some system for accepting household food scraps. Most provide educational opportunities to the public, either in the form of guided tours (62%) or composting workshops (56%).

The report occupies an interesting investigative/argumentative stance because of Louise's longstanding commitment to community compost. It's both an effort to figure out what exists *and* to explain to DSNY and other funding streams why community compost is important. Before joining DSNY, Louise had started Compost for Brooklyn in Kensington and worked with the Western Queens Compost Initiative, which later became the NYC Compost Project hosted by Big Reuse. Louise, referred to by one community composter as "our gal on the inside," offered an important perspective for the report because she had more experience at the community level than previous Compost Project coordinators. Louise explains that she and her successor, Marguerite, were the first Compost Project program managers to come from the community compost world, which signaled a shift in the department's attitude toward community compost.

In a climate where DSNY support for community composting comes under attack because it doesn't divert a significant chunk of the organic waste stream, the report explains how these projects steward green spaces, engage the community, and enable local organics recycling. The report acknowledges that community composting "diverts only a small amount of organic waste compared to a scaled-up NYC Organics Collection program" (DSNY 2014, 6). Instead, it goes into detail describing the benefits of compost to the city's "nutrient-depleted and sometimes contaminated urban soils" (11). The report explains how this relationship between community composting and soil health helps to support DSNY's mission of increased diversion: "They [community compost sites] raise awareness about what compost is and its benefits through outreach and education, and by using compost to grow food and care for green spaces in New York City neighborhoods. Making and using compost locally demonstrates firsthand to New Yorkers that apple cores and eggshells are not garbage—they are useful resources" (6). In addition to framing community compost as helpful for greening city property and for changing cultural values around organic waste, the report recommends several steps to "optimize the use of existing compost sites and resources" (43), including better funding, increased networking opportunities for community composters, better data collection at community compost sites, and leveraging community compost sites to improve outreach for DSNY's curbside organics collection. When other bureaus at DSNY started to take on language from the report, Louise was excited about "having them kind of copy and paste language, or seeing some of the language show up in OneNYC [the NYC sustainability plan]."

At Compost for Brooklyn (C4B), there are four clipboards with worksheets on them, and Nancy shows me how to record the number of people dropping off food scraps, the number of visitors, the number of volunteers, and the weight of food scraps being dropped off. Nancy counts everyone who stops by as a visitor, including a child in a stroller and the elderly man who stands outside the fence and hands little candies to me and Jeff. When a six-year-old mashes the chopper twice in a bucket of food scraps, Nancy counts him as a volunteer.

There is a binder in the shed at C4B in which Nancy keeps each day's worksheets. She explains that they compile them into an Excel spreadsheet for their yearly reports to their sponsoring nonprofit, and whenever they need to chase down grants. That Excel work is a volunteer shift that's great for anyone who doesn't like physical labor, she explains.

The Master Composter Manual makes the importance of data collection very clear: "If you can show your site is well-managed, you have more of a chance of securing funding" (4–39). The manual suggests recording the number of users dropping off food scraps, the number of volunteers helping out, the number of visitors to the site, and the weight of all food scraps collected. Many sites keep informal logs on-site—often in marbled composition books, crinkled with mud and moisture—and extract this data when needed for funding purposes. The Community Composting Report expresses dissatisfaction with the record-keeping practices of the sites it surveys: "Based on comments provided in the survey, it is clear that only a small percentage of sites have sophisticated systems for documenting the amount of organic waste they accept" (DSNY 2014, 13).

It's still not clear what sorts of data are useful, exactly. At the Big Reuse site under the Queensboro Bridge, the computerized aerator device records the temperature and oxygen levels of the first windrow. Leah explains that they're not quite sure what to do with this information: "It also gives us all these printouts, all this data of the temperature throughout cycles, which we haven't done anything with yet. But Charlie from Earth Matter was like, 'You should be analyzing it and sharing it with people you know' because, why not, we're at least gathering it, it's just not necessarily a priority at this point." Some sites in New York City use Farming Concrete, an online toolkit designed to measure the impact of urban farms on public health, city greening, waste diversion, and other measures. Resources such as the Master Composter Manual and the Community Composting Report recommend the website, and several large, visible organizations (for instance, Earth Matter and 596 Acres) use it to record their efforts.

However, the data at Farming Concrete is incomplete and often incommensurable. 596 Acres, for instance, only reports diverting material from the landfill in July, August, September, and December 2015, and records its diversion in July and December in pounds (100 pounds and 70 pounds) and its diversion in August and September in gallons (20 gallons and 10 gallons). Other gardens in New York City have similar problems with their data: it is sporadic, inconsistent, and often covers only a short range of time.

Microhaulers likewise run into challenges with data. At BK ROT there is a neon-green records notebook, weathered from long days outside, in which different workers and volunteers record the weight of collected food scraps and sifted compost each day. While Iyeshima's spreadsheets (about who paid how much and who left out how many bags) are flawless in Google Drive, the on-site records notebook is messy, and new tables are drawn up every day, using different methods for adding weights and taring the food scraps scale. It is very easy for food scraps to enter a compost system and for compost to leave the system without giving users any clear image of how much was added and how much compost was produced. I couldn't tell you what 20 pounds of food scraps looks like.

Work on citizen science has shown that grassroots data collection serves functions beyond simply generating numbers to crunch. Abby Kinchy, Kirk Jalbert, and Jessica Lyons have suggested that water quality monitoring can play a role in raising consciousness of environmental problems and fostering engagement (Jalbert and Kinchy 2015; Kinchy, Jalbert and Lyons 2014). For community composters, data serves a function as a site tries to understand its capacity for waste and its ability to produce compost, but this is not its primary role. Instead, data serves an outward role, legitimizing the existence of a compost project through its capacity to contribute to the city's understanding of its waste. The ability to generate data in and of itself becomes an important part of demonstrating the legitimacy of a compost system.

Matt Covert and Alfonso Morales describe how urban agriculture, once perceived as an informal activity and an undesirable, unregulated aspect of the urban economy, is transformed through reframing and regulation into a formal activity, and a productive aspect of a healthy community (2014). The Compost Project has performed a similar sort of reframing of community compost in New York City by reframing the informal economy of organic waste as an important part of local efforts to care for city soils, divert organic waste, and educate people about composting.

In 2010, many neighborhoods had gardens where you could compost your food scraps, there were a handful of farmers markets drop-off sites, and there were one or two businesses that could collect food scraps for composting. With the expansion of the Compost Project into supporting and organizing local compost, the project has made formal organizations out of several of the city's informal compost projects, and made the city's informal organic waste infrastructure searchable online. On DSNY's food scraps and yard waste collection page, you can select info on either curbside composting or drop-off; the drop-off option brings you to a map of compost drop-off sites across the city. Until the pandemic, the DSNY website also offered a list of compost sites, urban farms, demonstration sites, and community compost sites where you could "get involved or take a tour."

The Compost Project's efforts to unify community composters have succeeded in strengthening a network, providing recognizable signage for community compost sites, and making it easy to find local composting options. They have also resulted in several different representations of just what community composting is in New York City, and what it's capable of. However, in assembling this form of infrastructure, they also exclude projects that are community-oriented but charge for compost collection. The list of composting sites on the Compost Project page of the New York City website does not include sites like El Garden or Know Waste Lands, where the nonprofit microhauler BK ROT functions, and where organizers charge to pick up compost or accept drop-offs. It also excludes for-profit microhaulers like Vokashi or Common Grounds, which partner with community gardens.

While microhaulers have continued to be involved in the conversation, the emerging and crystallizing infrastructure for community compost has defined it as almost entirely not-for-profit, free of charge, and running on volunteer labor. This makes it difficult for small, paid compost services to gain footing in either the community compost field or the commercial compost field.

The COVID-19 pandemic posed a serious challenge to community composters. As part of the pandemic response, mayor Bill de Blasio and the city council cut $24.5 million from the DSNY budget. Of this, $7 million supported community compost, including funding for drop-off sites, local processing, and education. DSNY even reached out to community compost sites that it didn't fund but that were included in the city's list of drop-off sites, to demand that they stop accepting scraps in order to minimize contact risk.

In posts to the Community Composting Council Google Group, community composters tried to make sense of the funding cuts. Many people felt that the withdrawn funding for community composting was unjust, given the city's insistence on maintaining the $6 billion New York Police Department budget amid growing calls to defund the police. The order to close drop-off was also frustrating, and some argued that DSNY didn't have authority over community composters. Others defended the city's moves as necessary to slow the spread of the virus.

Compost advocates organized to address the funding cuts. Anna Sacks with Think Zero LLC, a zero-waste consulting firm, started a change.org petition: "Composting is essential to our environment and community. We cannot afford to regress." The petition eventually gathered over 20,000 signatures. The New York Lawyers for the Public Interest organized a

#SaveOurCompost virtual town hall on May 5, the day after organics collection was suspended. City council members, community composters, and microhaulers came together to demand that the city restore funding both to organics collection and to community composting. Over 1,000 people attended the Zoom event. The movement mobilized dozens of people to testify at city council budget hearings, and at hearings for two bills put forward to create organic waste drop-off sites in every community board. I write at greater length about the #SaveOurCompost campaign in chapter 2.

Microhaulers played a role in the campaign. In addition to speaking at events and testifying at city council meetings, they built online infrastructures for organizing. Common Ground Compost, a Manhattan-based bike microhauler, used its website to host the campaign's demands, share links to petitions and news coverage, and offer resources for home composting. And BK ROT created and managed a database of compost drop-off sites using Google Suite, which allowed people to find local composting opportunities in the absence of curbside collection.

Funding was restored for Compost Project host sites in the final budget that was approved in July 2020 (*BioCycle* 2020). However, this did not include funding for GrowNYC's food scraps drop-off at greenmarkets, which did not resume until September (Greenpointers 2020), without DSNY funding. And even once funding was restored, there were still threats to community composting infrastructure: The Compost Project hosted by Big Reuse and LESEC each faced possible eviction. Big Reuse had been operating its processing site on Parks Department land under the Queensboro Bridge, and Parks declined to renew the contract in order to use it as a parking lot. And the space in the East River Park that LESEC had been using for 34 years was threatened by a controversial climate adaptation plan that was meant to raise the entire park by eight feet. In both cases, composters felt that the Parks Department was not taking adequate action to find alternative sites for community composting. Again, the Save Our Compost Coalition mobilized to fight the evictions, and dozens of compost advocates testified at city council meetings, signed petitions, and contacted representatives to defend the compost sites. In June 2021, Big Reuse was able to renew its license with the Parks Department, and the LESEC received a written promise that it would be able to return to its site once construction was completed (Moran 2021).

COMMUNITIES OF COMPOSTERS

It is open hours on a Saturday at Compost for Brooklyn. An elderly Eastern European man with limited English comes in and hands me and Jeff candies from the Russian store down the street. We try to understand his one-word sentences and gestures until Nancy rescues us from the awkward interaction by greeting the man with love ("Boris!") and holding his hand while telling us stories about him in English.

After Boris leaves, Nancy stands at the garden gate, smiling and chopping scraps, and shouting "Good Shabbos!" at all the elderly women who walk by. She stands and talks with some of them—about Nancy's impending surgery, about their grown children, about the drops of rain that are falling and whether they'll turn into an actual shower. Being outside one or two days a week at a project like C4B (open, visible, on a well-traveled corner), and having the kind of personality that Nancy does (energetic, engaging) enable her to become what Jane Jacobs calls a "public character" for her neighborhood, helping to anchor a sense of shared community (1961). In a neighborhood like Kensington, which has a variety of immigrant communities that often don't interact with each other, Nancy's ability to bring people together in the green space of C4B provides an important public function.

Nancy beams with pride when she talks about the site's Muslim neighbors who volunteer at the site. Sometimes she offers a sort of casual ad hoc daycare at the garden, asking kids to chop food scraps and turn compost while moms and other caregivers run errands. She tells a story about a woman whose children spend time at the garden, who offered to be on call if ever Nancy needed help. She tells me how she worries about the women's hijabs getting stuck in the compost tumblers. The compost site has created a space where she can interact with neighbors she didn't know before.

She also glows about the adults with disabilities who come over from the Chabad around the corner once or twice a week. She tells me about wheelchair-accessible activities, like sifting compost or turning the auger in the curing piles.

From the early days, C4B tried to connect with neighbors. Louise tells me that they translated everything from the beginning, and their website offers instructions in Spanish, Chinese, French, Russian, and Polish.

C4B is not alone in its capacity to build makeshift communities around compost processing and the green space opened up by it. Other sites use food scraps drop-off times as a moment to open their doors and invite passersby to come help out or enjoy the garden.

"Community compost" is a slippery term. When Compost Project staff and volunteers talk about "community," they can refer to the inhabitants of a geographical region (as broad as a whole borough or as specific as a neighborhood), the people who have chosen to drop off food scraps at the compost site, informally organized people, or a public of stakeholders that are somehow affected by the project.

Marguerite explains that "it's really hard to define a community compost site because they're all different." When I spoke to her and Louise in the summer of 2015, as the Community Composting Report was getting published, both of them were focused on defining "community compost." When Compost Project staff set out to understand community composting, they found that self-identified community composters had different ideas about what makes something a "community" compost system: "Many sites have a lot of rules: no machinery, no electricity, everyone who donates needs to turn the pile, a variety of sizes and scales. Some focus on diversion, while other focus on education, or the production of food. This is an exciting variety!" When I asked why the Compost Project needed a definition of community compost, Marguerite explained that it was necessary for projects like the Community Composting Council or their site recognition program. Legally, the need to define "community compost" comes from the same anticorruption campaign that created the BIC—DSNY simply wants to be able to prevent big waste industry players from acting under the protected category of community compost.

For a site like C4B, it seems fairly easy to say "Yes, this is a community compost site." While the site is open only two to four hours per week, the site's facilitators understand their role as a green space in the neighborhood, and as stewards of their street corner. They create events that allow children to enjoy the space. They bring in adults with disabilities to help sift compost. Their finished product winds up in street tree beds. They interact with the neighborhood in a variety of ways, and the "community" that their site serves might be earnestly understood as the neighborhood itself.

When you go to Big Reuse's compost facility, though, the notion of "community" becomes more abstract. Tucked under the Queensboro Bridge, and hidden behind a Parks Department storage facility, the site doesn't have any of the inviting signage, greenery, or animals that a site like C4B or Earth Matter has. Just two neat, covered windrows, equipment and browns waiting to be used, and a port-a-potty. My only clue that there was a compost site there was the smell: strongly earthy, and a little piney from the Christmas trees being composted, but distinctly composty.[13]

But still, when I ask about it, both Leah and Ryan, the project coordinator in charge of food scrap drop offs, agree that Big Reuse is "totally a community compost site." For Leah this means that they are a resource for community gardeners, providing compost and expert advice. It also means that the future of compost at Big Reuse—the new site they were then designing—aims to be safer and more open to the public. Ryan reflected on the word "community" when we spoke: "This model of people coming to a drop-off site, taking ownership of something, knowing that today is drop-off day, 'I've collected my food scraps, I have to go and do this,' or even 'I want to go and do this'—that's a sense of that ownership, and I think that's a big part of what identifies or demarcates communities, is what do you own, what is your identity within the community. If you give people that rallying thing, that commonality, that's really big, that helps define community, so we try to encourage that." Ryan's statement identifies three elements that make Big Reuse "community compost": it gives people a sense of ownership, it offers shared structures, and it defines a common purpose for composting. When you drop off your food scraps with midscale compost operations, you might not be directly engaging with other composters, but you're acting together toward a shared goal. The "community" is all the people who feel ownership of Big Reuse, and find meaning in composting with them.

This version of community compost is useful for thinking about the small- and midscale fee-for-service models that also consider themselves community compost. A *BioCycle* article titled "The Business of Community Composting" profiles three community composters that charge for services, defining community compost as "the notion that organics are processed as close to the sources where they are generated to capture the benefits of both the process and the finished product for the community" (N. Clark 2015). At BK ROT, Sandy explains that one of their central goals is to build

13 The smell was probably sweeter in the winter than it was in the summer, when Leah told me that the neighbors in the public housing project on the other side of the tall chain-link fence would occasionally complain about odors.

infrastructure that can process waste in communities, and that communities are "willing to live side by side with." In these accounts, community compost is any compost system that considers the needs of the people who live with it.

Still, there are important differences between these imagined communities of food scraps drop-off and communities of people who work together to compost their waste. Ryan goes on to discuss how people come to identify as composters: "Technically these people aren't composters, they are saving their food scraps and they're bringing them to us. We're composters, you know, but in their minds, they think of themselves as composters, and so that's how we talk to people that are participants, they're composters. And in a sense that is true." For Gay Hawkins, the process of turning people into "recyclers" involves making people interact with waste differently. While she also argues that the kind of subject formed through recycling avoids considering the labor and environmental impact of the whole recycling system, she suggests that the embodied habits of separating recyclables from our waste is an important step in the production of an ecological consciousness (2006).

When I talked to P. about the early days of What a Waste!, they explained that they didn't just want to give people this shared identity as composters, they wanted to curate an experience of waste that would affect the people who brought in food scraps. They wanted to bring people to a central drop-off place to see it in the aggregate, and to do so in a social space designed to rupture the way people experience waste. P. recognized that this ran counter to hegemonic waste practices: "Giving it a form where it allows people to give a shit is maybe not always going to be acceptable to the status quo because it makes their jobs harder."

However, many people who want to separate out food scraps aren't looking for a disruption. Charlie tells me about a survey conducted by the Fort Greene Compost Project: "The first survey was three questions: Would you still do this, that is, come and drop your food scraps off, if you had to pay to do it? Everybody was okay with that. Would you still drop your food scraps off if you had to take them somewhere else, like, directly to the garden? Yes, especially if it was at a different time of the day or week, right. Would you still do this if you had to donate a morning every six months or so to help process it? Not so much [*laughs*] . . . thanks anyway, but no." Based on the passersby I've seen casually picking up choppers and pitchforks at community compost sites, I don't believe that the Fort Greene survey was indicative of a culture of food scrap separators in New York City that prefers being customers rather than community members. However, Charlie's results suggest that there's a limit to the kinds of subjectivities that composters are willing to take on.

Community compost is a nebulous category, and it means different things to different people. As a rule, though, it means compost infrastructure that is meaningful for a group of people—whether that includes customers, users, neighbors, or the people actively involved in the work. People use community composting services because they want to contribute to their communities. People do the labor of community composting because they want to be part of these projects. People live peacefully with community compost infrastructure because it is part of their social worlds.

Writing about community economies, J. K. Gibson-Graham challenge a prevailing understanding of community as a preexisting association of people that is always already inclusive and democratic. They write that "the association of community with a *being* that is already known precludes the *becoming* of new and as-yet unthought ways of being" (2006, 85). Instead, they describe community economy as "the active, ongoing negotiation of interdependence with all life forms, human and nonhuman" (Gibson-Graham et al. 2020, 5).

In some contexts, the "community" of community compost refers to preexisting associations: neighborhoods that need clean soil, green space, or good disposal options, or community groups that build their own composting systems. These communities are important for the Compost Project, because they demonstrate that community compost benefits people, and people are passionate about community composting.

In practice, though, community composters are just as interested in *becoming*: what new social worlds can we build through the work of composting together? How can these social worlds relate to their environment differently? Community compost systems are experimental infrastructures, open-ended in the kinds of systems and interactions they can facilitate. Both compost systems and the communities that inhabit them are emergent and unknown.

CONCLUSION: WHOSE UTOPIAS?

Hundreds of small- and midscale sites in New York City offer different models of community composting. Some invite volunteers to get involved in the labor of compost processing, while others minimize their demands on users.

Some facilitate the creation of community through face-to-face interaction, while others enable people to become part of a more abstract community of supporters. Some are vibrant parts of their neighborhoods, while others try not to make their presence known. Some receive funding from DSNY, and operate extensive programming, while others operate on a grassroots basis. These projects respond to local needs for green space, clean soil, gathering places, composting options, healthy outdoor activities, and more. For community composters, composting isn't just about waste diversion, it's part of a holistic approach to building community and space.

Neighborhood-based initiatives can change the ways people interact with each other and with their cities, meeting needs that are unmet by formal development (Anguelovski 2014; Moulaert and Swyngedouw 2013). Community compost sites embody and communicate local ideas about what urban sustainability can look like. They are experimental infrastructures that invite users to experience an alternative organizational principle for approaching the relationships between their households, their waste, their neighborhood, their city, and their world.

The NYC Compost Project puts these experimental infrastructures into a more complex relationship with neoliberal waste management. The Compost Project formalized some of the city's community compost infrastructure, and provided training and resources that have helped people build new compost sites and made composting available for more people. Support from the Compost Project has been shaped by people like Louise Bruce, who have brought their community composting values into DSNY, and tried to create an expansive definition of community compost. But infrastructuring community compost has involved decisions about what counts and doesn't count as "community compost." Policies and programs have pushed some specific approaches to community composting, to the exclusion of others.

One important dynamic is the exclusion of fee-for-service models. The Compost Project has favored a version of community composting that is free from financial interests, which means that community compost systems rely almost entirely on volunteer labor and are mostly unable to charge for products or services. This limitation is encouraged by the Master Composter Certification Course, funding relationships with demonstration sites, Parks Department regulations, and the Business Integrity Commission.

On the surface, these policies encourage the development of nonmarket systems for managing waste in community. But they need to be understood in the context of neoliberal waste management, which often calls on volunteer labor to meet the needs that are not met by market forces. Volunteerism asks people to perform important work out of the goodness of their hearts, instead of creating structures that can coordinate and carry out this work (Eliasoph 2013). In effect, the policies of DSNY, the Parks Department, and BIC set the rules of the community composting market in a way that discourages the development of alternative economies around organics recycling. This makes it hard to compensate compost labor outside the waste-oriented economics of large-scale commercial hauling. For the NYC Compost Project, community composting is an important part of urban sustainability, but allowing people to make a living off of this work has not been a priority.[14]

In its relationship with the Compost Project, community compost in New York City has elements of both abstract and concreate utopias. In my most paranoid reading, the Compost Project uses community compost sites as abstract utopias that mobilize idealized experiences of urban compost in order to support the less-than-ideal municipal curbside organics system. New Yorkers can witness the transformation of organic waste into compost, see how compost feeds the soil, and feel a sense of togetherness. But without a commitment to supporting these versions of the composting city, the Compost Project is just busy-ness, using community projects to kindle a kind of hope that DSNY is not committed to supporting. They are captured utopias: rooted in local dreams, and serving institutional goals.

I want to name this dynamic, but I don't believe it's worth pursuing as a total theory of compost in New York City. The Compost Project's continued support of these sites (in some cases materially, or through the production of compost experts; in other cases, only in educational materials and weak forms of advertisement) helps to create communities of composters that work together to build infrastructures that serve needs other than those of sanitation. Alison Kenner characterizes such a strategy as "civic infrastructuring," in which communities use the city's platforms to enact their own visions of engagement (2020).

14 I want to make it clear that I'm not advocating for a free-market approach to community composting. In New York City, free-market approaches to waste have historically led to racketeering, the unnecessary proliferation of waste infrastructure, and uneven facilities siting (Wood 2019; Stringer 2018; Reuter 1993). Instead, I want to point out that the exclusion of "dirty money" from community composting disadvantages alternative economic approaches, such as the model developed by BK ROT, which I'll discuss in greater detail in chapter 4.

Because they allow communities to imagine the city otherwise, community compost sites are also concrete utopias. They perform what José Esteban Muñoz describes as a utopian function: "The rejection of a here and now and an insistence on potentiality or concrete possibility for another world" (2009, 1). Site visits, volunteer days, compost drop-offs, and other interactions with community compost projects allow residents to engage with a future that is not yet here. Stepping off the street and into a community compost site feels like stepping into another world: people are friendlier, the city is quieter, and sometimes there are chickens. Even just dropping off food scraps at the farmers market or signing up for a microhauling service allows people to experience waste diversion as something meaningful that enriches their community—not just a virtuous deed.

Through these experiences, community compost systems allow designers, users, and visitors to inhabit a world in which waste infrastructure functions differently. They anchor people in alternative systems through new disposal habits, new chores, new kinds of community work. They invite people to get involved in the future-oriented work of building, maintaining, rotting, planting, and planning—processes that take months or years to complete. Bins and trees and sheds blend volunteer-run projects into timelines that extend past certainty in neighborhoods where lots might get bought up by developers. Community composting builds new futures that people can work toward together, where waste systems support communities.

The Compost Project treats community composting both as an educational project to support the transition to municipal organics recycling *and* as the foundation of an alternative waste system. With the city's investment in curbside organics, more resources have been put toward developing community composting sites as abstract utopias—projects that are meant to bring in volunteers and train people to separate scraps, but not meant to grow and support themselves. But the Compost Project could reorient toward supporting these utopian projects concretely, helping community composters actualize their closed-loop dreams *and* that further the institutionalization of organics recycling.

What would this look like in practice? City agencies could expand their definition of community compost to include fee-for-service models that center community needs. New York City could protect existing compost sites and grant unused land to community composters. The Compost Project could compensate more of the work already being done by community composters: hauling, processing, education, and technical support. And DSNY could involve community composters in decisions about municipal organics, to make use of their experience building organics recycling systems that fit meaningfully into New Yorker's lives. Such interventions could allow community composting to grow in surprising new ways, and articulate new visions for urban compost. Understanding community composting not as a tool in the transition to the standard model of organics recycling in New York City but as its own emergent infrastructure allows for the creation of more meaningful, surprising systems.

CHAPTER FOUR

BK ROT

COMPOSTING NEOLIBERAL WASTE MANAGEMENT TO MODEL JUST SUSTAINABILITY

The idea behind BK ROT is this: youth-powered compost pickup and processing in community gardens that creates well-paying, expert jobs and centers youth of color in environmental stewardship. Starting in 2013, the organization worked with four youth to collect compost from 45 households and two drop-off sites in Bushwick, Brooklyn, and to process it at two community gardens. The youth provide the labor for collection and processing, handle customer interactions, and represent the organization at public events. Wages have always been above $15 an hour, even before the Fight for $15 movement succeeded in raising the minimum wage in New York City in 2016.

Sandy Nurse came up with the idea for a youth-of-color-run bike-based compost pickup service while she was delivering food by bike in 2011. She started shopping the idea around with activists, gardeners, and compost enthusiasts, and eventually linked up with Renée Peperone at a 596 Acres meeting.[1] The two of them piloted the project, with Sandy doing pickup from ten households that they found through community garden circles, and in 2013 they were able to hire R., Iyeshima, D., and T.,[2] students at the Ecostation program at Bushwick High School. After processing their

1 596 Acres is a community gardening organization in New York City that helps gardeners to get access to empty lots.

2 Due to limitations in my Institutional Review Board protocol, I was not allowed to formally interview anyone under 18. I had many conversations with youth workers, but I don't quote them directly. I'm using pseudonyms for all youth workers who were under 18 at the time of fieldwork. Unfortunately, this means that the story in this chapter centers on organizers rather than youth workers.

materials at Bushwick Abbey for a year, they began working with El Garden, a new garden space on Jefferson St that wanted a source of compost. In 2014, they acquired the rights to a lot on Myrtle Avenue at Dekalb Avenue and teamed up with like-minded gardeners to build a wildlife garden and compost site. They eventually named the new site Know Waste Lands.[3]

Sandy is a carpenter by training, and carpentry is how she moves through the world. She builds solid things, including physical structures, relationships, systems, and movements. Since starting BK ROT, Sandy ran a successful campaign for city council, and has become a fairly recognizable figure in New York City politics. In the most widely used campaign photo, she gazes upward hopefully. She is Afro-Latina, and her hair floats in a cloud around her face. She wears a yellow sweater, bright red lipstick, and gold hoop earrings. Her mouth is straight, but she smiles slightly with her eyes. This is part of the Sandy I know—the visionary builder of systems. But the image fails to capture her openness, her ability to see eye to eye with people and to find places for connection and play.

Renée is a perfect foil for Sandy. She is white and petite, with bright eyes and an energy that you can always feel simmering below the surface. She is the daughter of a New York City waste management family, returned to the family business in a way, but probably not the way anyone expected. She speaks in conspiratorial tones; when we talk, I often feel like she's sharing some private secret.

BK ROT incorporated as a nonprofit in 2015, and hired Sandy as part-time executive director in 2018. In 2019, Sandy and Renée both stepped away from the project to pursue other goals—Sandy to run for city council, and Renée to focus on her career as a documentary filmmaker. They hired Ceci Pineda, who served as executive director from 2019 to 2021. In 2020, the project composted 73 tons of compost, employed seven full- or part-time youth workers and four youth leaders, and paid out $86,000 in wages to youth workers.

BK ROT is an experimental infrastructure for composting, modeled after the principles of environmental justice. As a working model of a possible

3 In June 2015, Sandy and Renée were still playing around with the name Wasteland or The Wasteland for the garden but getting some pushback from the board on the grounds that it's culturally insensitive to call a place a "wasteland" in Bushwick, where the discourse and aesthetic of emptiness and possibility made Bushwick so appealing to developers and "creatives" in the first wave of its gentrification. I suggested the playful No Waste Lands, which didn't get much traction at the time, but would tie together the zero waste ideologies that a compost project appeals to and push against the idea that Bushwick had been a wasteland before it gentrified. The current Know Waste Lands takes these two ideas but the pun on Know/No suggests that the place is designed to help people get to know waste, as well as the "wastelands" in which the project is taking place. The Spanish name Tierra Sin Desecho (Land without Waste) loses the puns on "knowledge" and "wastelands" but offers another, global implication: "Earth without waste."

FIGURE 4. Know Waste Lands, 2016. Photo by author.

future, the project invites publics to witness and take part in a composting system that poses radical challenges to the standard model of organics recycling. The project demonstrates two key innovations in waste and space: a reliance on expert, self-determined, well-compensated, visible labor, and the use of waste infrastructure to facilitate accessible green space in a gentrifying neighborhood.

This model responds to a variety of problems in existing formal and informal waste infrastructures, including the often-exclusive market for green jobs (Taylor 2014; Torres-Velez 2011), the often-exploitive market for waste jobs (Pellow 2003; Alexander and Reno 2012), and the exclusivity of the largely volunteer labor pools responsible for much of the city's community compost infrastructure. The BK ROT vision for public space runs counter to both the historical overburdening of Bushwick (and North Brooklyn more broadly) with waste transfer and disposal sites (Wood 2019; Corburn 2005), and the redefinition of Bushwick's public spaces through gentrification (Zukin and Braslow 2011). BK ROT organizers have been particularly sensitive to the problem of *green gentrification*, a process in which sustainability initiatives contribute to gentrification (Gould and Lewis 2016).

BK ROT works within a *just sustainabilities* paradigm that combines environmental and social justice goals. Julian Agyeman describes just sustainability as a "bridge" between the historically white "new environmental paradigm," which prioritizes environmental sustainability over social justice, and the environmental justice paradigm, in which scholars and activists have largely focused on the uneven, racialized distribution of environmental bads, rather than environmental goods. In a just sustainability paradigm, activists pursue justice and equity within ecosystem limits, striving to improve quality of life for both future and present generations (2008). Isabelle Anguelovski describes how communities facing environmental injustice go beyond fighting against pollution to pursue just sustainability, creating food systems, community spaces, good housing, and economic sustainability (2014). Writers such as Nik Heynen, Laura Pulido, and Juan De Lara have also described a tradition of abolitionist ecology or radical ecology, in which activists build ecologies that challenge not only the histories of environmental justice but also the larger histories of racial capitalism, colonialism, and white supremacy (Heynen 2020; Pulido and De Lara 2018). For BK ROT, compost is not only a solution to waste problems, it's part of a holistic approach to environmental and social justice.

Other microhaulers (small-scale compost haulers) have pursued compost for social justice goals. Worker-owned hauling collectives such as Denver Compost Collective in Colorado, and Hard Core Compost in Santa Cruz, California, challenge the traditional economic models of waste management. Some worker-owned projects, such as CERO Cooperative in Dorchester, Massachusetts, and GreenFeen OrganiX in the Bronx, New York, respond to racism in the waste system and green economy by cultivating compost systems led by Black and Brown people. In Maryland, the Baltimore Compost Collective operates in protest of a garbage incinerator in a predominantly Black neighborhood: the collective diverts organics from the waste stream in order to "starve the beast" of incineration, while also plugging resources back into the local economy (Fabricant 2022; Aning 2021). And Veteran Compost in Aberdeen, Maryland, responds to high unemployment rates among combat veterans with employment opportunities for veterans and their families.

BK ROT has made use of new markets for organics collection to meet its social justice goals, but this has not always been an easy task. Alison Alkon's study of a farmers market in the predominantly Black neighborhood of West Oakland demonstrates that even in "green for all" projects that aim to empower Black farmers, retailers, and consumers, participation in a neoliberal market can lead activists and entrepreneurs to prioritize economic rather than social justice goals (2012). Working in the green economy of composting means that BK ROT has had to consciously push back against the dominant frames of neoliberal waste management and gentrification, which include the taken-for-grantedness of waste labor, exploitive economic models for waste businesses, and the availability of urban space for development. Under BK ROT's fee-for-service model, the project's leaders have been troubled to find that the majority of its customers are white and middle class. However, BK ROT's focus on labor, its movement-oriented leaders, and its reliance on grant funding have helped to center the project around its principles, rather than on the promise of growth.

Projects like BK ROT critique the depoliticized approach to sustainability transitions, in which a successful transition simply reduces the environmental impacts of present systems. Organics recycling can unfold in a variety of ways, responding to present social, cultural, and political circumstances, and building new ways of inhabiting the world. Sustainability transition policy and research need to recognize the existence of diverse approaches to sustainability, which might challenge more than just the environmental impacts of current systems.

BK ROT's designers and maintainers have been involved in an ongoing process of bringing the radical idea of compost for just sustainability into conversation with the social and material forces of waste, urban development, markets, and nonprofit logistics. In this chapter, I describe some of the tensions that BK ROT has navigated in trying to build a people-centered model of compost in a neoliberal waste market. I use the metaphor of composting to describe how BK ROT has met and responded to these tensions, building an alternative system that responds to neoliberalism, gentrification, and environmental injustice. Laura Pulido and Juan De Lara describe this as an important part of radical ecology, which recognizes the "potentiality for radical futures that respond to the shackles of the present but that are not confined by them" (2018, 5).

The accounts that inform this chapter draw mostly on a three-year time span from 2014 to 2017, when I was working with BK ROT in person on a weekly basis, sometimes more. It is also informed by my ongoing involvement with BK ROT from 2017 to 2022, as a member of the advisory board and board of directors, and as board chair. One outcome of this relationship is that I've foregrounded connection over ethnographic detachment in my account of BK ROT.

TACTICAL INCLUSIVITY IN GENTRIFYING BUSHWICK

The gentrification of Bushwick has been similar to the gentrification of Williamsburg, the Lower East Side before it, and similar districts and cities elsewhere. As artists were priced out of the previous creative district (Williamsburg), they began to move further out on the L train to the next neighborhood, where rent was still affordable for working artists. Bushwick in the 2000s was a racially mixed, mostly Latinx neighborhood, with significant industry and some relatively noxious waste transfer facilities. As "creatives" moved in, a combination of new art spaces and boutique food stores made the neighborhood more appealing for people with middle-class incomes (Zukin and Braslow 2011). Bushwick's street art, abandoned buildings, and taco trucks offered an "authenticity" that was marketed to potential residents (Osman 2012; Greenberg 2008). With the influx of middle-class residents, landlords were able to increase rents. This made it difficult for the working-class, majority Latinx, often immigrant residents who had lived there previously, as well as for the artists who had just moved in, to afford to stay. During this study, Bushwick was at a moment in its gentrification where much of it was still affordable for working-class people, but it had a well-recognized cultural capital and was the site of significant development

(Zukin and Braslow 2011). The average monthly rent for a one-bedroom apartment in Bushwick rose 29.8% between 2014 and 2015 (J. G. Keil 2014). Kathe Newman and Elvin Wyly put the displacement rate from gentrification between 6.1% and 9.9%, or 10,000 displacements per year (2006).

What is happening in Bushwick is part of what Jason Hackworth and Neil Smith call the "third wave of gentrification" (2001). This wave has moved beyond Manhattan and is generally led by developers. Unlike in previous waves, antigentrification organizers have been largely unable to rely on city support. The Department of Housing Preservation and Development, which once cheaply sold abandoned buildings to housing advocates, now offers housing vouchers and requires that luxury buildings include "affordable" housing (Newman and Wyly 2006). The city has also relaxed its zoning laws and reduced protection of affordable housing (Hackworth and Smith 2001).

In addition to economic shifts, gentrification has come with more direct regulatory interventions that have directly targeted Black and Latinx people in Bushwick. The selective application of sanitary regulations at neighborhood bodegas has been used to weaken a market that is largely Latinx-owned (Herrera 2015). And increased policing in order to "clean the streets" (Cahill 2007) made these areas dangerous for Black and Latinx people, particularly young men.

Sandy has been involved in movements against racialized police violence since before she started BK ROT. She wore a "justice for Akai Gurley" patch on her shirt the first time I met her.[4] She describes the role of Immigration and Customs Enforcement as part of the gentrification of Bushwick: "And you think about the fact that policing is used as a tool to clear neighborhoods and meet quotas. . . . once a week you'd have a checkpoint over here and they'd just be pulling people over for the secure communities ICE stings. . . . And then like lo and behold you have Troutman and Jefferson Street and Starr just like popping off as this new Williamsburg scene so it was this really systematic—let's wipe this whole area out and start this beautiful natural foods store and all these clubs and bars for people that aren't from here, so in my mind it's all connected." Activists and scholars influenced by Black Lives Matter have emphasized the relationships between police violence and environmental injustice (Mercado 2020; Linnitt 2020; Pellow

4 Akai Gurley was a Black man who was fatally shot by a police officer patrolling a public housing stairwell in 2014. Gurley was unarmed and had not confronted the officers.

2016). Lindsey Dillon and Julie Sze argue that the dominant, market-based approaches to sustainability will always tend toward surveillance, gentrification, and overpolicing, and conclude that the movement against police violence requires an alternative approach to sustainability, one that brings a race and class analysis to the question of which environments should be sustained (2018).

Gentrification has also involved a cultural shift that has made Bushwick more hospitable to the desired new residents (artists, white people, middle-class people) and less hospitable to people who already lived there. Ali Madanipour explains that cultural markers and regulations work together to change the cityscape for different people: "A combination of formalized rules and regulations, informal codes and signs, and fears and desires control our spatial behavior and alert us to the limitations on our access. Through these, we have come to know whether we can enter a place, are welcomed in another and excluded from others. More restrictions on our access to our surroundings would bring about the feeling of being trapped, alienated and excluded from our social space" (2015, 162)." Even as developers marketed Bushwick's "authenticity," new cultural institutions, such as performance venues and coffee shops, and new street art, often commissioned (Zukin and Braslow 2011), helped to give it the feel of an "arts district."

BK ROT organizers have been sensitive to the conflicting roles that community gardens can play in gentrification. A variety of authors argue that community gardens can be empowering spaces for marginalized people, foster a sense of community, and ensure a "right to the city" for people who are often left out of urban planning (Anguelovski 2014; Armstrong 2000). However, an emerging body of literature describes the ways that community gardens contribute to neoliberal retrenchment and "green gentrification." By tasking city dwellers with the creation and maintenance of public green spaces, community gardens can justify the withdrawal of support for urban parks (McClintock 2014; Ghose and Pettygrove 2014). Gardens in low-rent neighborhoods can also attract developers who use green spaces as a selling point, and affluent renters who garden as a hobby (Gould and Lewis 2016; Checker 2011; Quastel 2009).

During a lull in work one day while I sat and chatted with Iyeshima, a white woman in a pashmina and a North Face jacket entered the garden and sat with us to drink her latte. She had just moved to the neighborhood, she said. *I can tell*, I thought—she didn't look like the people who had lived here for decades, or like the people who moved into the artist spaces in Bushwick ten years ago, or even like the hipsters with marketing jobs who had moved here in the last three years. The space was beautiful, she said, she's so glad this is so close to her house.

Kenneth Gould and Tammy Lewis describe how developers in Brooklyn encouraged the greening of Prospect Park, Brooklyn Bridge Park, and the Gowanus Canal as part of a conscious effort to drive up real estate prices in local neighborhoods (2016). The High Line, a public park built on an abandoned railroad track, has had a widely criticized impact on rent and culture in the Chelsea neighborhood of Manhattan (Loughran 2014; J. Moss 2012). "Green infrastructure" projects, which use urban green space to create climate resilience in developing neighborhoods, can raise rents and price out poor residents and residents of color, pushing them into areas that are more prone to climate disaster (Anguelovski et al. 2019).

It's difficult to escape awareness of gentrification in Brooklyn, where many neighborhoods have condos going up or a new pie shop moving in. And it's difficult to escape guilt about it. Most of the artists, students, hipsters, and queer people in my circles are aware of and feel guilty and joke about our role in the gentrification of our neighborhoods. Even though we're able to afford only modest rents, we make a neighborhood look more appealing to middle-class renters and developers (Schulman 2012), and we can't help but support the combination bar/flower shop that just moved in three blocks away. Many of us are practiced in guilt, and our default attitude toward gentrification is sardonic resignation, coupled with an effort to differentiate ourselves from "*those* gentrifiers." Renée, a white woman from Westchester who moved to Bushwick 20 years ago, laughs at the graffiti tag on a new housing complex that says, succinctly, "white people." "It's true!" she says as we both pause to snap photos on our phones.

Awareness and widespread guilt around gentrification have created a market for finger-pointing—at developers, store owners, and street artists who are the visible, tangible perpetrators of urban change. Intrusions like a Wes Anderson crochet mural in Bushwick (Whitford 2015), an upscale cereal bar in London (Lawless 2015), or a pop-up restaurant on the roof of a recently shuttered trade school in Philadelphia have attracted anger and protests directed at very small pieces of a much larger problem (Otterbein 2015). Sarah Goodyear argues that the obvious faces of gentrification are a poor target for activist energies: "By directing rage about seismic economic and social forces at such puny targets, people who care about urban neighborhoods risk dissipating their energy and letting the parties who are truly responsible for displacement and disinvestment get off easy" (2015). Kristyn Stewart, an education activist writing on misplaced gentrification anxieties in Philadelphia, suggests that policymakers who defund education systems and allow unchecked development and rent increases are a more effective and deserving target for the energies directed at cultural producers (2015).

Antigentrification activism has often intervened in the cultural mechanics of gentrification. In these strategies, activists and organizers subvert the meanings attached to artifacts designed to attract gentrifying residents and developers. Strategies such as "just green enough" environmental cleanup function to make spaces healthy and livable without making them "quaint" (Curran and Hamilton 2012). Jennifer Wolch, Jason Byrne, and Joshua Newell suggest that "just green enough" strategies might include focusing on weaving small amounts of green space into existing neighborhoods, planning for functional green spaces, and combining urban greening with antigentrification activism (2014).[5] With the site at Know Waste Lands (KWL), BK ROT makes a cultural intervention in gentrification by building public green space that is meant to be open not just to gentrifiers but also to the working-class Latinx and Black people who lived in Bushwick before gentrification started. Murals reflect the experiences of long-term residents, signs are in Spanish and English, and Latin American culinary herbs such as epazote and *pápalo* grow in the garden.

Renée explains to me how making KWL as open as possible to the Black and Latinx residents of Bushwick often means it has to be strategically closed in some ways. Planning takes place among a restricted group of actors. Renée explains that "every single person that was kind of gravitating to be in the garden community was a certain demographic, and we didn't feel like it was reflecting the neighborhood." Renée explains the tactical exclusion that went into designing KWL: "People think around community gardens that they have to be like fully like community based . . . everybody should have input, but actually . . . when you're in a community that's gentrifying, that means that the people who have the time and energy the most,

5 While there's some interesting work that's been done around "just green enough," the phrase leaves a bad taste in my mouth—the apparent implication is that Black and Brown people and working-class people don't need as much "green" as wealthier, whiter residents. There's nothing less "green"—either literally or figuratively—about urban greening projects led by people of color. BK ROT is simply interested in a less marketable shade of green.

and the means, . . . are gonna be the ones determining the space, right? So it's like, not only are you dealing with race and class, but you also have to actually hold back a little bit and process over time, because not everybody has shown up yet, right?" (E. K. Foster 2015). Anguelovski explains that community green spaces are sometimes meant to function as "borders," demonstrating that a space "belongs to residents—and not to outsiders" (2014, 27). Closing down the design process for the garden was used as a way to design the space to be more open to Latinx and Black people, working-class residents, and long-term residents.

This exclusion extended to me as well. When I first started coming to work with and write about BK ROT, Sandy and Renée took a few weeks to warm up to me and talk to me about the thought that went into their project. They asked probing questions to try to understand why I was interested, and I found myself going out of my way to demonstrate that my political beliefs generally aligned with theirs. Renée explained that they're nervous about other people taking the "bike compost delivery" part of their idea without the "good green jobs for Black and Brown youth" part.

BK ROT organizers moved their operations to KWL in the summer of 2015, and invited neighbors to an open house in September. The new lot is right underneath a train station, in the acute angle formed by two well-traveled streets. Not only is it in a highly trafficked area, but it provides a tempting shortcut: you can see right through it. Sandy, Renée, and Hilary publicized the event by distributing fliers in grocery stores and bodegas in both English and Spanish, and by talking with community leaders about the new space. And a team of mural painters—Raul and Fernanda—had been designing a mural for the space, informed by interviews with neighbors who had lived on the block since before Bushwick started gentrifying.

The KWL party made the space feel open. People walked in off the street, and the resulting crowd was mostly Brown-skinned and Spanish-speaking. When Renée announced that a "ceremony" was about to start, I shuddered, thinking about the neopagan rituals that white garden members used to practice at El Garden while D. and I tried not to notice. But the ritual that took place was apparently legible and inviting to Latinx visitors. An older *espiritista* in a white dress tied with a red sash burned herbs in a cup while men beat drums behind her. They told the assembled crowd to turn to each of the cardinal directions and repeat phrases. While Iyeshima, A., Hilary, and I all half-complied, the rest of the crowd was more enthusiastic.

FIGURE 5. The group Cetiliztli Nauhcampa performs a ritual at the Know Waste Lands open house, 2015. Photo by Alex Seel. Image courtesy of the Laundromat Project.

I'm wary of trying to code this ceremony as authentic, given the cultural cachet attached to authenticity in gentrifying Bushwick—not to mention in ethnography. It was less alienating than the El Garden ceremonies, and commanded respect, or at least engagement.

After the ceremony, Iyeshima and A. told visitors about the role of BK ROT in the garden, and people lingered to talk to them about compost, treating them as the experts they were. We served empanadas from a local shop, apple cider, and hot coffee. Sandy taught A. how to make s'mores. A story in the *Brooklyn Paper* called the space "the prettiest dump in town" (Hobbs 2015).

BALANCING ODOR AND JUSTICE

My first few times volunteering with BK ROT, I turned the compost piles with W., a Black 16-year-old youth worker who had been dragged into the organization by his friend, D. While D. had been on the "good green jobs" scene for some time now—he had already worked with Green Guerillas for a year [6]—W. was new to the world of environmental work. I think W. joined either to hang out with D. or to make some extra money. Unlike D.

6 Green Guerillas is a gardening organization that provides material and political support to urban gardens and farms, and also runs a youth employment and education program.

or Iyeshima, it was hard to get W. talking. The Sundays when it was just me and him, we worked mostly in silence. I tried to get him to talk in my usual avuncular way ("How's school going?"), but he gave monosyllabic responses to most of my questions. We would shovel the compost out of each bin and onto a tarp, then shovel it back in, mixing in the week's new food scraps along the way, and covering the pile with a thin layer of sawdust.

Smells were one of the few things we ever talked about. A compost pile can smell horrible, even when it's turned regularly and cared for well. This is because the populations of bacteria in the pile change as the pile rots, heats, and condenses.

I should explain some of what's going on at the microscopic level here. There are two broad classes of bacteria that are involved in composting (alongside a wide variety of fungi, protozoans, rotifers, and arthropods): aerobic and anaerobic. Aerobic bacteria require oxygen, and create carbon dioxide as a byproduct. Anaerobic bacteria don't require oxygen, and give off methane as a byproduct, along with smelly compounds such as hydrogen sulfide and ammonia. Both sets of bacteria will turn organic materials into humus, but aerobic bacteria will work faster and generate heat that can kill pathogens. The two sets of organisms are usually working side by side; to say that a pile is "aerobic" or "anaerobic" means only that a specific kind of bacteria is dominating at the moment. An aerobic pile will usually have some anaerobic activity going on, producing unpleasant odors, as well as other organisms such as actinomycetes that create a characteristic earthy smell. Introducing more oxygen to a pile, either by turning it or by using a static aeration system, helps to make the environment more hospitable to the aerobic bacteria. But as low-oxygen regions form in clumps of moisture, even a well-turned pile can develop pockets of anaerobic activity that hold the sulfur and ammonia smells that we associate with rotting food (Cornell Waste Management Institute 1996).

Each of these pockets smells different. Some you can enjoy—a little whiff of artisanal fermentation, like bleu cheese or kimchi. But others can slap you in the face. Turning a compost pile with someone means breaching little pockets of anaerobic activity and finding surprising new smells together. W. accurately described one of these smells as "baby poop," and we could use that as a label for a specific, unforgiveable smell. Sharp, always strong, distinctly fecal. The baby poop smell was the one that I'd always get stuck on my boots and drag around with me to coffee shops after working at BK ROT. W. and D. were smart to put plastic bags over their high-tops to keep the smell from sticking. The first time W. called the smell "baby poop" I laughed out loud, it was so dead-on, and he laughed along with me. After that, just saying the phrase "baby poop smell" was enough to get him to smile.

Smells provide a bonding experience for the people at BK ROT—we have to deal with horrible smells on a regular basis, and we laugh at them together while squishing up our noses. When I ask Hilary, a perennial volunteer, why she keeps coming back each week, she explains that "doing this kind of boring and unpleasant thing together is bonding in a weird way . . . like you're all my buds, from doing this."

This isn't a project of reveling in gross smells; BK ROT has always tried to keep the piles aerobic out of respect for the neighborhood. Smell has been a source of anxiety for Sandy and Renée, who are aware of the fraught history of smell as a component of waste infrastructure. But in their early days, despite the team's best efforts, turning the pile would sometimes send the sharp smell of anaerobic activity beyond the chain link fence at El Garden and down the street. Some Sundays, I could detect it half a block away.

Even though odor isn't a health hazard on its own, it carries a symbolic weight, especially in communities like Bushwick that are already overburdened with waste infrastructure. Carlos Pedilla, speaking for the South Bronx Clean Air Coalition, suggests that foul odors place an emotional burden on communities: "You begin to understand that you are less" (quoted in Sze 2006). And the smells that linger on waste laborers have long functioned to mark them as dirty or less than others (Nagle 2013).

For BK ROT, smell is part of a balancing act between two different approaches to environmental justice (EJ) that have ethical and practical implications. The first I'll label a "NIBBBY" approach: "not in Black and Brown people's backyards," following Robert Bullard and Beverly Hendrix Wright's articulation of the "PIBBY" principle (put in Blacks' backyards). PIBBY is the standard industry response to "NIMBY" (not in my backyard) activism: when white activists reject locally unwanted land uses such as toxic waste facilities, those facilities often end up sited in Black and Brown neighborhoods, which have less representation and fewer resources to push back (1987). The NIBBBY principle suggests that the proper place for waste infrastructure is anywhere *but* Black and Brown communities that are already overburdened. In this perspective, building any waste infrastructure in Bushwick is problematic. Sandy and Renée were sympathetic to neighbors' concerns about odors—and neighbors *would* complain at El

Garden, occasionally loudly, or just with scrunched up noses as they pass by the fence. Sandy and Renée knew it was important to make sure they're making no excessive demands on the olfactory systems of the surrounding community.

The second approach is a critical environmental justice articulation of zero waste, which recognizes the ways that racism and colonialism are not only products of geographical decisions about the fate of wastes but are also baked into the ways waste is generated and managed (Pellow 2016; Liboiron 2021). Sandy argues that the "out of sight, out of mind" attitude of waste management in New York City has helped to create the uneven burden of waste infrastructure (and its odors) because it has preferentially routed waste through Black and Brown neighborhoods on its way out of the city. "We need to learn how to live with waste," she sighs when she hears about controversies over curbside organics collection in Greenpoint (see chapter 2). "Living with waste" is not just about settling in to the uneasy and unhealthy cohabitation forced by traditional waste systems—residents learning to deal with waste infrastructure as it's been designed.[7] Instead, it's about learning to build waste infrastructure that is worth living with. It involves what Nik Heynen characterizes as "re-Earthing," which involves "a re-configuration of the power, land and outcomes of a people and the world they live in" (2020, 11). Living with waste demands new ways of sharing labor and space.

BK ROT balances the NIBBBY and zero waste approaches to waste infrastructure, building a system that works to keep waste present while trying to make sure there's a difference between "presence" and "dump." Sandy explains that they work to counter the idea of Bushwick as a dumping ground: "Waste as a literal and symbolic dump in Black and Brown neighborhoods is absolutely not lost on me." But she also reminds me that the stuff they deal with is not toxic: "It's not industrial waste infrastructure that has all these toxic outputs. This is actually pretty healthy waste infrastructure. It's good for everything." Certainly, the occasional sharp baby poop smell that floats down the street from BK ROT has nothing on the lingering sour milk smells of nearby waste transfer facilities, with their deep notes of mechanical grime.

7 I want to address a possible ecocolonizing reading of "we need to learn how to live with waste." When I showed this section to Sandy, the quote unsettled her, since it seemed to echo the "you will learn to love it" of so many projects conducted in the name of urban greening. When Sandy shared this feeling, I took it as a statement of a politics of waste that diffracts a classic zero waste ideology (there is no "away"; we will necessarily live with waste) through the radical world building that is possible in projects like BK ROT (we, collectively, need to learn new modes of living). For me the statement "We need to learn how to live with waste" encoded an ethics of world-building that reincorporates waste into the world.

Sandy and Renée have worked tirelessly to manage the olfactory presence of BK ROT; they've done their homework, checking in with neighbors and negotiating ways to offset their odor with free compost. Over time, they've learned how to better manage their piles to keep odors down.

I got a call from Renée after I e-mailed her a draft of this section. "Compost doesn't smell," she insisted. "If you manage it right." She spoke quickly, connecting dots between ecological and socioeconomic systems with the wide-eyed intensity I'd come to expect from her. Smell had triggered something. "The problem is that we don't have the resources or the labor to manage our compost to the point where it doesn't smell." Sandy had also mentioned this problem: up until 2016, BK ROT just didn't have the expertise or the resources to manage a compost pile without odors.

Renée stepped back and compared BK ROT's work to the kind of composting that permaculturists do. She had come to BK ROT through permaculture, which uses compost to close a loop between organic wastes and soil fertility. "Permaculture is about balance," she explained. This balance is pretty easy to manage when you're just looking at a farm or a homestead, but that's not what BK ROT is working with: "We're part of a larger system, a race and class system, that's out of balance. And we're expected to create balance in this system."

The baby poop smell is what happens when BK ROT tries to balance its browns and its greens while simultaneously balancing its waste collection and its compost sales, its organic methods with its social justice values, and local waste infrastructure with respect for its neighbors.

At the time of writing, the smells of BK ROT have been under control for a while. In spring 2016, BK ROT hired Domingo as a lead composter. He and I joke about his nuclear reactor heart—the man explodes with an energy that seems supernatural to the rest of the team. Most of it is directed at compost, too: he works six days a week at various compost sites in the city, including Added Value Farm in Red Hook and Brooklyn Botanic Garden. He quickly took charge and made changes that allowed the system to flow faster, handle more material, and smell better. The secret was more water, surprisingly—he showed us how to water the piles until leachate flows out, which makes them get hotter and allows microbes to work faster.

The addition of Domingo marked a transition in the shape and smell of the compost system: it became far more palatable to outsiders once he got all the youth workers to chop the scraps, add enough browns, and water the piles enough. The new system is not perfectly smell-free, and rats are sometimes a problem, but odor problems are minimal.

I walked into KWL one afternoon in 2016 and found the crew working. Domingo greeted me jovially, leaning on his big shovel like Fred Astaire leaning on an umbrella. "Do you smell that?" he asked. Once I reached the concrete pad, the smell hit me. But it was far from the baby poop smell of El Garden's piles, or the unholy kimchi smell of bokashi systems, or the stale smell of day-old trash bags. It was just an active smell: distinctly garbage, a little sour, but already halfway to earth. "That means it's alive!"

KALE AND THE POLITICS OF URBAN GARDENING

By June 2015 I'd gotten into a schedule with BK ROT, biking to El Garden in Bushwick every Sunday around 1 p.m. for a few hours of smelly, arduous labor. Hilary compares this routine to church: a place you go on Sundays that's altogether not quite enjoyable but makes you feel better about yourself and the world. When I showed up on June 21, I was on my way to compost church, prepared—and dressed—for physical labor and the smells of compost-gone-anaerobic. But when I turned the corner onto Jefferson Street, I heard live music playing, and when I walked into the garden I saw 20 to 30 people (mostly white) chatting, and there was a tent set up near the compost bins. I locked my bike to a sign outside, and went in to try to figure out what was going on, and if it made sense for me to be there.

I didn't see Renée or Sandy, but Hilary was chatting with her friends by the snack table. She explained that the garden was having a picnic. She didn't share my sense of disruption, and I tried to chat with her for a minute before excusing myself to grab a plate and check out the snack table.

W. and A.—two of the youth workers—were eyeing the table as well, but their plates were empty. I didn't blame them. The centerpiece was a stately steel bowl of raw, chopped kale—not a massaged kale salad, but simply kale. To the side was an undressed mix of cold lima beans and chickpeas. To drink there was a dispenser labeled "lemonade" that held lemon-flavored water. A white woman saw the kids' empty plates and asked if they were hungry. W. shook his head. "Do you want some kale?" she asked.

The bowl of kale was unforgiveable to me. Kale has been much-lampooned as a gentrifier's food. My favorite depiction is in the "Settlers of Brooklyn" video, a parody of Settlers of Cataan, in which kale features as a resource with which settlers can build out their colonies into new areas of Brooklyn that are "virtually uninhabited by young adults with wealthy parents" (Bierbaum 2015). Kale isn't *only* a tool of gentrification: it's healthy; it grows easily in all sorts of weather; and has been prepared in countless ways by a wide variety of cultural groups. These leaves are thick with meaning. However, watching a white woman offer raw kale to Black high schoolers demonstrated that this picnic was not designed with W. and A. in mind.

I see this a lot in my fieldwork: people who work in gardens trying to convince young Black men to eat raw unseasoned vegetables. The culinary missionaries fit into what Julie Guthman calls the "if they only knew" framework of organic advocates: if people who don't love vegetables could only eat a fresh-picked tomato, they'd see how much they love and need fresh food (2008a). The problem with this is that tastes are learned, and experiences of food are not inherent in that food. My love of freshly picked tomatoes is shaped by my upper-middle-class background, a childhood diet based mostly on *Moosewood* and *Silver Palate* cookbooks, and my ongoing addiction to the Cooking section of the *New York Times*.

It's an oversimplification to say that food simply serves as a cultural barrier. Like anything else growing in El Garden, kale is full of possibility. While M. turned the compost one Sunday in November, Renée worked on getting a garden bed ready for the winter. She talked with me about M., how quickly she turned the compost, how graceful her tiny figure was as she moved these huge, stinking piles of dirt. Renée picked some cosmos, a fennel blossom, and a baby leaf of kale and tied them together with twine. When M. finished turning, Renée called her over and handed her the bouquet and said to put it in water once she got home. The gesture was loving, and M. responded with sincere appreciation.

While much of the policy and academic discourse around community gardens has explored their food provisioning aspects as a source of resilience (Shimpo, Wesener, and McWilliam 2019; Chan, DuBois, and Tidball et 2015; Okvat and Zautra 2014) and as a tool for community building (Firth et al. 2011; Flachs 2010; Armstrong 2000), other authors point out that the social and cultural complexities of food make it a difficult point of engagement for building community. Julie Guthman argues that the food provisioning efforts of community gardeners serve only to re-create white cultural values (2008b; 2008c). Rachel Slocum argues that despite their good intentions, alternative food practices often carry with them a whiteness and an exclusivity (2006).

If food is useful for building a community at El Garden,[8] that community is largely made up of white transplants rather than long-term residents of color. BK ROT's insistence on creating a community garden space—Know Waste Lands—devoted to waste, rather than food, creates an alternate set of dynamics around the community compost bin.

BUILDING COMMUNITY WITH WASTE

One of BK ROT's first advisory board members was Ana de Luco, a Spanish nun who founded and manages the Sure We Can cooperative recycling center in Bushwick. Sure We Can is a place for collectors to store and redeem bottles and cans—a place that supports the strong culture of scavenging in New York City. Sure We Can and other recyclable collectors have come under fire from DSNY for "robbing" the department of potential profits from recyclables (Mencel 2014). But Ana fights with a steely certainty. For her, scavenging offers a socially useful kind of work that is available to homeless people and people living with mental illness and addiction.

Sure We Can resembles the community recycling centers of the 1960s and 1970s, which also used waste labor as a way to create jobs for dispossessed people and to build connections between waste generators and waste workers (Lounsbury, Ventresca, and Hirsch 2003; Pellow 2003). Following this tradition, BK ROT uses waste and waste labor as a way to create good jobs for Black and Brown youth who are often excluded from green economy jobs, and to build inclusive and open community space.

While marginalized groups are often forced to live with dirty, unsafe, or degrading waste systems, many people have also found ways to repurpose wastes. A variety of scholars and storytellers have described the creative appropriation of wastes by groups who are denied access to other resources (Potter, Osseo-Asare, and M'Rithaa 2019; Kendall 2007; Varda 2000). However, the role of waste as resource often creates a cultural mark on people (Nagle 2013), subjects them to dangerous conditions (Amankwa 2014; Pellow 2003), and justifies austerity (Vasintjan 2015). In neoliberal waste

8 El Garden was created with help from 596 Acres, a network that provides tools for New York City residents to secure land for community gardens. Renée associates 596 Acres with "the new guard" of community gardeners, which has separated from the radical roots of community gardens. In the 1970s and 1980s, the community gardens were projects that responded to the city's failure to create safe spaces and green spaces for residents (Schmelzkopf 1995). The 596 lots, according to Renée, don't meet local needs as much as they provide new spaces for hobbyists. "You can use their organizing tools to actually get land," Renée explains, "but that doesn't mean you know how to organize a community."

management, the idea that waste is a resource can justify removing support structures if marginalized people can rely on the revenues available in wastes that need to get taken care of anyway.

An alternative approach to waste-as-resource involves the strategic use of the cultural associations of waste as part of a movement strategy (Schaffer 2014). In this approach, the abject status of waste is used to further movement goals. Garbage workers' strikes are an excellent example of this (Fredericks 2014): the disgust of trash becomes a weapon in the hands of garbage workers fighting for better working conditions. Similar strategies have been taken by the "trash" T-shirt campaign, which uses "trash" as a metaphor for the invisibility of queer and trans people of color (williams 2015) or the documentary *The Throwaways*, the title of which references the director's disposal at the hands of several social support systems (McKinley and Suchak 2013).

In their efforts to create jobs and green space, BK ROT organizers treat food scraps as both a resource and as waste. The abjection of food waste figures in their advertising: their name uses the word "rot" rather than a more positive term such as "food cycles" or "green solutions." A comic strip that Hilary produced to explain the service makes prominent use of smell lines and gross-out faces. At a fundraiser in 2016, Sandy, Renée, Hilary, and I made decorations that were meant to resemble rotting food and leachate. Treating waste as gross helps BK ROT frame the service as desirable for customers, and to establish the labor as difficult and deserving of compensation. While BK ROT does not revel in disgust—rotting food smells are still an environmental justice issue—organizers deploy disgust in certain contexts to convey the importance of this work.

Waste management also offers some affordances for community building that are not offered by community gardening. BK ROT's status as a waste disposal site helps to open up the space: while food gardeners need to develop a relationship with a space that is time-intensive and ongoing, disposers can stop by to throw out their scraps casually. A man who raises pigeons on a rooftop across the street (a source of anxiety for some of the workers, who worry about getting pooped on by the circling flock) walked through the open garden to deliver some pigeons to a neighbor, and on the way back struck up a conversation with Renée that ended with him offering his pigeons' accumulated droppings for the compost piles.

FIGURE 6. Hilary's BK ROT comic (Allison 2015). Image courtesy of Hilary Allison.

The waste labor taking place at Know Waste Lands also makes it available to more people. As a Green Thumb garden,[9] KWL is required to offer 20 open hours per week, and while other members initially scrambled to commit to times, the garden could immediately count on at least five hours from BK ROT just because of the regularity of their Sunday workdays. The garden is inviting when people are working in it: people stop by, strike up conversations, and leave their kids to do garden tours with BK ROT organizers. It reminds me of Jane Jacobs's description of sidewalks as safe spaces for children because of the watchful presence of their "natural proprietors" (1961).

Finally, compost specifically has a certain "science class" quality that organizers deploy when younger kids wander into the garden. "Want to see some worms?" we ask them, and then put them on the imaginary task of pulling worms out of the compost that we're sifting. Renée says that kids as old as high schoolers will walk into the garden and get excited to see how compost works.

IMAGINING FUTURES TOGETHER

I met Nando at the Community Composting Council meeting in February 2016. He works up in Harlem at a project called Brotherhood Sister Sol, which provides job training, maintains a community garden, and was exploring bike-based food scraps collection. He came to BK ROT twice in March and April to talk about a partnership with his organization. Each visit ended in an excited huddle, talking about new visions for local compost in New York City, or job training programs that could create a corps of compost experts (not just haulers) working locally. He sat down with the BK ROT youth workers and talked to them about developing a worker-owned collective ("Do you know what that means?") that could work alongside DSNY without becoming incorporated into them. He and the BK ROT organizers, and Domingo and myself, talked about a future in which trained and organized compost technicians were in charge of the city's organic waste, and made sure that small compost facilities were running smoothly, taking care of waste, and doing good in their neighborhoods. R. and A. listened attentively, asking questions. The conversation went on with the same quiet intensity, and it was hard to leave.

At BK ROT—and similar projects—I've found people engaging in casual collective dreaming as a day-to-day activity. This happens a lot around

9 The garden is on Parks Department land, which makes it subject to certain city regulations.

waste: we find some potentially useful resource and try to imagine how to plug it back into a system of use and reuse. I brought this up with Renée, and she suggested that the dreaming helps to create a sense of community, like we can work together to design some alternative future that we can agree on—even if that future is just a dream. The feelings of possibility that we get when we share in the excitement of future making help to create the sense of community that we experience at BK ROT.

The youth workers haven't always been part of this excitement, but Sandy and Renée encourage them to get more involved. In fall of 2015, Sandy and Renée started having monthly meetings before Sunday workdays that they could use to check in with the youth. Unsurprisingly, three high schoolers on a Sunday morning had a hard time talking, much less talking about compost. Check-ins were monosyllabic. But at the November meeting, when Sandy mentioned a new lot she was trying to acquire to turn into a farm with actual livestock, A. and R. started awake. They had both immigrated from rural areas in Latin America, and grew up near animals. When Renée asked the youth about their goals with regards to BK ROT, R. said (with a rare smile) that he wanted to start a farm, with chickens and goats, and eventually cows. Sandy laughed and said cows might be hard, and R. backed down but remained insistent on the chicken idea. "They would give us so many things! Eggs, poop, meat." I'd never seen him that animated. Renée (vegetarian) and Sandy (vegan) both cringed at the idea of using the chickens for meat, but A. laughed and offered to snap some chicken necks. Renée was shocked at this—she thought she was seeing some bloodthirsty side of a kid she'd come to trust. The moment of excitement about an urban agriculture future brought out something we didn't always get to see in those kids.

Iddo Tavory and Nina Eliasoph describe future orientations as central to social life: "Actors have to share an image of a future together, even if implicitly—at least enough to coordinate action, if not necessarily enough to coordinate thought or desire" (2013). But Renée cautions me that these dreams often fail to materialize, and that an abundance of dreams can be the sign of a frustrating project, when people do the casual dreaming thing without creating alternative systems.

In contrast, she brings up the mural that Raul and Fernanda are designing at KWL, which weaves together residents' own concerns about police brutality and gentrification with the promise of a space that is designed to be open to Central and South American immigrants. The piece brings together the future orientation of the BK ROT and KWL coordinators with the histories and imaginations of its neighbors. The mural also pins down the space as a specifically immigrant-oriented garden project. Murals, of course, are open to appropriation by developers as a source of "authenticity" in neighborhoods like Bushwick (Fitzmahan 2013), but a well-made mural can create a sense of place and belonging (Chakravarty and Chan 2016; Telhan et al. 2014).

While Sandy and Hilary and I (but mostly Sandy, who is a trained carpenter) built the compost bins at KWL, Raul and Fernanda brought in people they met on the street—mostly brown-skinned people, Spanish speakers—and sat them down on stools in the center of the garden to interview them. They asked about their relationships with the neighborhood, what they wanted to see in the space, the changes they saw as new people moved in.

Raul and Fernanda are a mural painting team, and they had been commissioned (through the Laundromat Project[10]) to paint murals for the new garden. They wanted to ground their murals in the experiences and dreams of long-time residents, as a means of making it clear whose space the garden was. Raul has long hair drawn back into a ponytail, thick rectangular glasses, and tattoos all up and down his arms. When I met him, he was wearing a heavy metal band T-shirt. He seemed shy at first but quickly opened up when I talked to him about his work. Fernanda is warm, full of smiles, affable to a degree that amazes me. She wears vintage buttoned dresses and big earrings. Both are from Ecuador.

I approached Raul after one of their interviews, and asked him if I could interview him about the work they were doing there. He laughed and said he had wanted to interview me, at which I shrugged and waved my hand and explained that I wasn't from Bushwick at all, that I'd just moved to Brooklyn, that I might even be temporary (I was). He agreed, and said that my visions weren't the idea behind the mural.

The mural project is titled "Whose Streets?" and the team sent out preliminary drawings in mid-November, which were followed by enthusiastic e-mails from the board of KWL. For the fence facing Dekalb Avenue, they designed four heads with plants growing on top of them: three women and one man, all people of color. One looked like the *espiritista* who had led the ceremony at the open house; the others I didn't recognize. The plants

10 The Laundromat Project is an arts nonprofit that aims to put socially relevant art in "laundromats and other everyday spaces" in order to open people up to new ideas and connect them with their neighbors (The Laundromat Project 2023).

(bouquets of amaranth, bee balm, and goldenrod, all native to the Americas) sat on their heads like crowns. For the Myrtle Avenue side they drew an urban cityscape in the background with broken concrete in the foreground, the "wasteland" of the garden's name. In the center, a man, a woman, an elder, and a child surround a plant in the soil, sheltered from the wasteland by oversized leaves. When Raul sent out the drawings, he asked the team for suggestions on native birds. He'd been thinking about pigeons,[11] but another team member suggested hummingbirds to evoke the migratory lives of the people who had lived in Bushwick, for whom the garden was being built.[12]

The mural is an effort to put Black and Latinx faces on a garden space that can easily be read as white. The images tie the garden in with the life stories and cultural touchstones of people who have lived in Bushwick for 30 years. Raul explained that he intended to include images of increased policing and the violence of development in further sections of the mural, alongside images of the social life of the neighborhood before gentrification, and the possibility of rebuilding around green spaces, but these sections have not yet been completed.

Murals have been used as a means of tying a community to a space and forming identity (LaWare 1998), and visualizing problems that might otherwise be invisible (Corburn 2005). Raul and Fernanda's practice suggests another use of murals, as a form of community-based design. These pieces not only respond to the interests of neighbors (in a community being consciously constructed by the organizers), they provide images of what kinds of potential the space holds. Carl DiSalvo describes how the field of speculative design uses tracing and projection to investigate and make visible the forces that shape our technological landscapes (tracing) and imagines what the world might look like as these forces continue to act in the future (projection) (2009). Raul and Fernanda's work traces the histories of the built environment—and the destroyed environment—in Bushwick, but also

11 Pigeons are the unofficial city bird of New York City.

12 KWL was billed as a "native plant garden," but in the handout Renée gave me to help identify the vegetation there, fewer than half of the plants listed were native to New York. The framework of "nativeness" can cut in multiple directions when looking at the politics of living in Bushwick: Bushwick "natives" can be understood as the victims of a colonizing outside force. At the same time, contemporary anti-immigrant discourse often frames Latinx immigrants as the dangerous outsiders. The figure of the "invasive species" has come under fire in environmental management in recent years because it relies on xenophobic or racist ideas about "foreign invaders" (the Mexican "killer" bee, for instance) (Raffles 2011, smiley 2014). By framing the garden as "native" and planting immigrant plants, Renée demonstrates that the concept of "native" isn't natural and puts it up for debate.

the work that is taking place in community spaces in order to project an alternative vision of the neighborhood's future. They are concrete utopias, connecting this vision to the everyday work of gardening and composting.

I FINISHED THE dissertation that turned into this book in 2016, and I've now had six years to watch how imagined futures have materialized and dematerialized. The farm dream has been tabled, but youth workers have gotten more involved in shaping the direction of BK ROT. Ceci Pineda replaced Sandy as executive director in 2019, and has prioritized creating space for youth leadership: "As much as possible when we're talking about policies or creating new things for the organization," they said, they want to make "space for youth voices and listen to their ideas and try to figure out how to implement them."

Ceci has overseen the development of the Youth Leaders Program, a summer employment program that trains youth in composting, environmental justice, and advocacy. Youth leaders have developed relationships with partner gardens in which they provide technical assistance to their community compost systems. Ceci has also supported youth workers in moving into leadership roles and engaging in advocacy. And while Nando's discussion of a worker co-op didn't create immediate change, the board of directors and leadership team started researching worker co-op models in 2021 as a way to give youth workers more control over the organization, and also to distribute the responsibilities of management.

Youth workers have also brought their own ideas to the table to shape the direction of BK ROT. Starting in November 2018, R. (by this point, the operations manager) and S. (then the operations specialist) drew up a plan for their own residential organics project. BK ROT had transitioned entirely over to commercial collection in 2017, because it was more efficient and because the city's curbside organics program threatened to make their residential pickup service obsolete. However, when DSNY halted the expansion of curbside organics in 2018, R. and S. recognized that a market still existed for residential organics collection and decided to coordinate and operate a residential pickup service on their own. The residential pickup service could also create more work hours for bikers. They called the service Rabbit Roots, a nod to the rabbits that R. still dreamed about. During the pandemic, when curbside organics collection was cancelled and many drop-off sites were put on hold, Rabbit Roots was able to expand to meet the new demand for residential collection.

POURING THE FOUNDATION FOR UTOPIA

In May 2015, I went to Know Waste Lands to "help" lay the concrete foundation for their new compost system. Renée and Sandy and I tried to lend a hand, but the men pouring the concrete got annoyed by our efforts, and shooed us out of the area. Or they put us on smaller, easier tasks, like holding boards in place ("I think nails can do this" suggested Renée). They didn't want us screwing up this process: once the concrete was out of the mixer, it was already starting to harden, and the smallest mistake would be essentially permanent. The men moved quickly and forcefully. They shoved the heavy, sludgy stuff around with shovels, following a builder's logic that I couldn't understand, and that they couldn't explain to us. They tried to show us how to use a two-by-four to even out the top, but Renée and I kept creating ripples in the surface, and the men came back and fixed it.

Once the initial hubbub died down, though, and most of the workers had come to rest in the shade, Ronnie (the head of the work crew) grabbed the hand of one of his workers, stood on the wooden edge of the slab, and leaned out over the concrete, his colleague leaning back to counterbalance him. Ronnie ran a moistened trowel over the surface, smoothing it out. Renée and Sandy and I stood under a tree, marveling at how these men who had been so fast and frustrated an hour ago had entered a meditative ballet once the action settled down.

The concrete pad was the result of crowdsourcing; $1,205 came from 13 donors to buy the materials.[13] Ronnie volunteered his crew's labor. For Sandy and Renée, the concrete pad provided a much more practical space for doing compost than the woodchips they used to work on. They blamed the Brooklyn topsoil for contaminating the organization's first batch of

13 Sandy and Renée had tried for months to find extra concrete—concrete that had been mixed on accident, in excess, or in not-quite-the-right proportions—hoping that they could make use of waste concrete for free or cheap. In the end, it was too difficult to coordinate the volunteer labor they were relying on for laying the concrete pad, and the excess concrete that they were hoping to come by. Perhaps one of the biggest problems with economies that rely on excess is that the excesses rarely match up with each other: extra concrete rarely coincides with volunteer labor.

compost with lead and other heavy metals.[14] Concrete would be an easier surface to work on, and would provide a steady platform for the compost bins and other equipment. And it would keep rats from burrowing into the compost piles from the ground. But there's also meaning work going into concrete: unlike the plants, wooden structures, and tools that made up the rest of the group's infrastructure, a concrete pad put a semipermanent, location-specific function on the space. While BK ROT had already created changes in culture, environment, and material flows using tools and plants, the concrete pad was a kind of technological legislation (Winner 1980) that didn't necessarily mean that the garden would be a youth-powered community-minded compost site for the foreseeable future, but it made a mark that was sure to last.

But concrete is not usually celebrated for its environmental benefits. It creates runoff and limits plant growth, which is particularly problematic in urban spaces, where exposed soil is hard to come by. And it lasts a long time—one friend who was also a master composter expressed concern about pouring a concrete pad for a project that might not be around forever. Would the potential good of BK ROT really outweigh the environmental impact of concrete?

My friend's reading of BK ROT made me worry about my commitment to them, both as a researcher and as an activist. Was BK ROT the kind of project that could fall apart if its organizers lost steam? And what would it leave in its place if the compost stopped turning? I've been involved in organizing projects that ran on pure optimism, and I've seen them fail when they relied too heavily on volunteer labor and donations. I've learned, through a long sequence of quiet disasters throughout my life, that grassroots efforts often *do* fall by the wayside when life gets too complicated. And at the time, I had heard both Sandy and Renée express feeling overextended, and wishing they could step back a bit from running BK ROT.

For my friend, the concrete pad represented the harm that the project could do if it fell apart. But this reading of the structure robs it of its meaning to BK ROT. The concrete carried with it a sense of hope and *concreteness* that carried the project into an imagined future. When Sandy built the four-bin system on it, it glowed with promise. A year later, with the pad mostly full of composting equipment, Sandy, Renée, and Domingo would

14 The health of Brooklyn topsoil is an important question in the creation of compost systems in New York City.

stare at it and excitedly try to Tetris another windrow into it. Three years later, tall windrows would fill up every square inch of the concrete pad, and the mountain ranges of decomposing material would get hot enough to send up steam in the winter. Five years later, BK ROT would break ground on another site two doors down, and the board of directors would try to imagine which of our wildest dreams would look best on it—a composting toilet? A biogas stove?

José Esteban Muñoz describes a similar glow in Kevin McCarty's photographs of the stages at punk clubs and gay bars. He calls this feeling *anticipatory illumination*: "That moment of possibility right before an amazing band or performance manifests itself onstage and transforms the world for the performance's duration and, for many of those in attendance, beyond" (2009, 104). Muñoz uses a play on the word "stage" to get at this relationship between anticipation, performance, and transformation—there are the stages in McCarty's photographs, but also the "it's just a stage" that parents imagine their queer children to be in—a stage that, for most, never ends. Muñoz turns the "just a stage" of queerness on its head: what if we take queerness to be "just a stage" in the performative sense, with all the excitement and transformation and potentiality that's built into these structures? The "stage" becomes an important metaphor for understanding the queer version of utopia that Muñoz describes: a specific location that promises something transformative, and a temporal container for something that extends into the future.

Returning to the concrete reality of KWL, Muñoz's reading of stages reframes my friend's concerns about the long-term impacts of a potentially short-term passion project. Rather than a future ruin, we can read the concrete pad as a stage that provides a present container for a potential future. The concrete pad makes the work of composting more sustainable in the present: it helps BK ROT workers produce cleaner compost, maintain a more sanitary system, and move materials around more efficiently. It also provides a stage on which BK ROT can perform something hopeful, promising, and transformative: an approach to organics recycling that is built to uplift the people and places that are most burdened by existing waste systems. The semipermanence of concrete connects present efforts to an alternative future, framing the maintenance work of shoveling scraps and caring for compost as part of a long-term vision. It allows BK ROT to function as a (literally) concrete utopia, anchored in the here and now but offering new possibilities for the not-yet-here.

FIGURE 7. The new bins on the concrete slab, 2015. Photo by author.

VALUING WASTE LABOR

In those early, informal days, I would trip over terms when I talked about the youth who spent their Sundays biking food scraps around, turning piles and answering e-mails for/as BK ROT. When I wrote that "BK ROT employs four youth workers" in a draft of a press release, Renée corrected me in a comment: "Engages? Technically they are not 'workers', we cannot employ, so maybe we need to shift language a little." In the final draft, we used the suitably vague "BK ROT works with four youth," which evades the question of who is working for whom.

Youth employment is a central concept for BK ROT. "Good green jobs" are why the organization was founded in the first place. David J. Hess explains that the definition of a good green job is up for debate, but that a "good job" generally means a job that has "benefits, family-supporting wages, and potential for career development" (2012, 54), and a "green job" is often taken to be work in a "business that engages in some kind of green activity" (63). These kinds of "good jobs" are difficult to create in a grassroots organization like BK ROT, where nearly all staff members work part-time because they're in high school.

Sandy has taken a more holistic approach to "good green jobs." She cites Grace Lee Boggs's "new work" philosophy as an inspiration. Frithjof Bergmann, a disciple of Boggs, describes new work as meaningful, self-directed, and community-oriented (2019). Ceci offers their definition of a "good green job," which likewise builds on the new work philosophy: "I think it is something that is honoring workers, it honors the land or our relationships to the earth. And it is also not contributing to suffering . . . for other people, or places or beings." On top of more traditional definitions of a "good green job," BK ROT also pursues the creation of jobs that are self-directed and that put people in right relationship with community and environment.

Youth employment brings new challenges to the good green jobs project—not just legally but also interpersonally and in terms of getting work done. A large part of the work being done at BK ROT is navigating a way through relationships with youth workers, who alternately act like employees, students, family, coworkers, and friends.

For most of my time volunteering regularly, the work was split up between three youth workers and Iyeshima, who was in college at the time. On a typical Sunday in 2016, R. and A. would do the pick-ups, M. would turn the compost, and Iyeshima would send out emails to members and manage routes and payments from a nearby coffee shop. The day would start with Iyeshima or Renée posting print-outs of the routes for R. and A. on the door of the shed at El Garden, then at noon they would show up, pick up their bikes and trailers, and spend two hours collecting food scraps.

Biking in Brooklyn can be dangerous, so organizers have tried to minimize risk to workers. They've brought in trainers such as The Brown Bike Girl to teach basic safety for bikers. They've developed relationships with local bike shops in order to keep bikes tuned up. BK ROT offers free helmets to bikers, courtesy of Nutcase Helmets, and free cold-weather gear, courtesy of Patagonia. High schoolers don't always want to wear helmets, so there have been occasional fights with Sandy and Renée over the expectation to wear a helmet while biking.[15] Despite the training, bikers will still be careless sometimes, which makes all the adults worry. There have never been any serious accidents or injuries.

Some days, R. would drop off his little brother, H. (age six). Depending on the weather, A. and R. would stop in the middle of the day to drink some bottled water (provided by Sandy) and sit on the benches by the shed, in

15 Nutcase wasted us to post pictures and videos of bikers wearing helmets on social media, which created an extra level of awkwardness for already-shy high schoolers.

FIGURE 8. Victor poses with bike trailer at El Garden, 2015. Photo by Renée Peperone.

the shade. A. would chat with me and M. about the pickups, R. would sit in stony silence or entertain H. if he was around. When they came back, they would weigh the food scraps they had collected and put them in a plastic tumbler to be chopped up by M. A. would bring the material from the Bushwick Food Co-op to KWL, where he would chop it up on his own, mix it with browns, and add it to the bin.

M. managed the food scraps that were already at El Garden. She would take the temperature of each compost bin with a long thermometer that she poked deep into the pile, and write it down in a notebook so that we knew which piles were active. She would lay down a tarp and use a pitchfork to pull most of the first bin's compost out, then chop up last weeks' scraps from the tumbler, mix them with sawdust, and add them back to the first bin with the rest of the decomposing materials. She would pull out the next bin's compost and return it to the bin. She would switch between pitchfork, flat shovel, and rounded shovel while she worked, depending on the texture of the material and the surfaces she was working on. She would work with earbuds in, listening to music. Renée would sometimes stand back and marvel at the way that M. moved hundreds of pounds of stinking, rotting food. Sandy would beam as she told me what a great image M. is for Black girls—not just a Black girl composter but a pretty, popular, cheerleader Black girl composter who could deal with smells and hard work while still maintaining impeccable hair.

M. would get lonely turning the compost alone—A. used to work with her, but since the project is growing, he started spending most of his time picking up compost and processing it at the other garden. I would work with M. if I had time, and talk to her about school, food, books. One week she told me excitedly that she got to see a human brain in her neuroscience class.

D., who had been in charge of turning compost before M. showed up, would put plastic bags over his sneakers to keep them from getting soiled. M. followed suit. When a pair of youth from Brotherhood Sister Sol came to learn about the BK ROT system, Sandy showed them the plastic bag trick and they picked it up.

Iyeshima would hang out at Express Yourself Café a block away, waiting for customers or workers to contact her with issues. Her work wasn't limited to Sundays—she was on call for customer questions all week, and would answer them within 24 hours. Between managing the spreadsheets, mapping out the routes, answering e-mails, contacting members, and responding to new concerns, she clocked ten hours of work with BK ROT per week on top of her work as a college student. Iyeshima explained that her job was to "try to be that medium where I don't confuse the members with what we're doing behind the scenes of everything"—her work was about making sure the changes going on in the organization (new work schedules, new employees, new site) didn't create disruptive changes for members. If Sandy and Renée were both out of town, Iyeshima was the de facto leader. She enjoyed this responsibility: "It's great that I'm in charge. I feel like in the beginning, I should have been in charge." She has always had a confidence that amazes me.

The adults would hang out with H. whenever he was around. His and R.'s parents worked six days a week, so leaving him at home for three to four hours on Sunday afternoons wasn't an option. Hilary and I took babysitting as a fun new challenge. Neither of us was used to being around kids, so we'd run around and play games with him, creating space in which R. could go out and work, and Sandy and Renée could tend to the things they needed to do. If the adults were doing something simple, like sifting finished compost, we would put H. on worm duty: picking worms out of the sifter and putting them back in the pile. The kid was photogenic—three missing teeth, thick glasses, and an easy smile. His efforts to help out frequently showed up on our website: H. spraying a hose, H. holding a worm. Sandy worried a little about using him as a mascot.

Hilary and I enjoyed hanging out with H., and it seemed fair that R.'s job should come with daycare service. But there were weekends when none of us had the energy or focus to play with H. Like many six-year-olds, he wanted attention and could be a little exhausting. Renée started asking R. not to bring his little brother around, and he usually agreed not to. I'm not sure what H. would do when he wasn't there.

The youth workers mean a lot to the adults who work at BK ROT. Hilary lists them as a reason for coming back every week. For Sandy and Renée, the kids took on familial roles: Sandy explains that R. is "like a little brother" to her, and Renée brought her "aunt skills into it in some weird way." For the white members of the team, including myself, they were an important site for interracial/intergenerational connection. Hilary explains that they're kids she "would not know at all, you know," if it weren't for BK ROT, and Renée says that they let her see "a world I'm not privy to in any other way." I regularly tried to pry into the kids' lives to try to figure out what's cool among high schoolers right now. I never learned much.

Youth work is the focus of BK ROT, but it involves a lot of labor on the part of the organizers and volunteers to make sure that youth work goes smoothly. When Sandy and Renée were designing BK ROT in 2012, they ran a one-month pilot in which Sandy picked up food scraps by bike from 15 houses, just to make sure the concept was workable in practice before they brought in youth workers. Sandy and Renée were almost always present on workdays, helping to direct workers and volunteers, managing issues as they arose, and stepping in to do the work themselves when needed.[16]

Part of the challenge for Sandy and Renée was creating a system that self-regulates. Sandy explains that "it's kind of like starting to create these roles where the youth can actually take it on and step up and we can step back in a way that isn't damaging. Like we're just handing something to them but it's like skill development and personal growth, professional growth." Both coordinators would dream of the system someday managing itself: "I really want to get to a point where Renée and I are on the board, you know?"

Over the three years that I came to workdays regularly, Sandy and Renée's involvement moved from being right there working alongside the youth to going between the two gardens and coordinating things in the background, a little less present than they were before. In November 2015, I asked Renée what she did for BK ROT. She paused, searched her memories, and replied, "I think that the kids are taking on a lot of work and I'm sort of like . . . I'm not sure exactly what I do." That didn't mean they were working less, though—Sandy and Renée's work had shifted from operations to management—handling paperwork, grants, a nonprofit structure, partnerships, and accounts—creating an organizational container that kept the physical infrastructure operational.

Still, the physical system required frequent monitoring by Sandy and Renée to make sure everything was working smoothly. On a few occasions, Renée would check on the workers' activities and realize that their compost game wasn't up to par. For a few weeks, M. had switched the order of the compost piles, putting fresh food scraps into the second pile instead of the first. I had asked her if she was sure about this, but it took Renée's certainty to convince her that she had the order backward. At KWL, A. wasn't adding nearly enough browns for Renée's tastes, so she showed him how to make sure the whole pile was covered in sawdust to make it less appealing to flies.

16 For most of its existence, the "good green jobs" that BK ROT was trying to create referred only to the youth and young adults who biked, turned compost, and coordinated routes—the day-to-day work of operations. The behind-the-scenes work of holding this infrastructure together—managing workers, pinch-hitting when workers were out, writing grants, organizing fundraisers, giving tours of the site, filing taxes, hiring and firing workers, fixing hardware issues, communicating with neighbors, speaking at events, and many other small and large tasks—was still almost entirely volunteer labor. Sandy and Renée had the support of a community, but they were taking on the vast majority of this labor, along with all the work of coordinating volunteers. Even though BK ROT was created to compensate the work of environmental stewardship, stewarding the environmental stewards was all-volunteer for the first five years. This changed in 2018, when BK ROT found funding to hire Sandy as a part-time executive director. Sandy continued to do much of the same work she had always done—essentially working full-time—but now she made a salary and had a title.

"Making sure that the kids, even though they're doing it every Sunday, that they're maintaining the structure of what we laid out a little more formally, because it gets loose if we're not . . . I think it starts to unravel if you don't show up to bark a little bit." This barking doesn't sit well with Renée, who doesn't enjoy being a rule maker. "It feels kinda shitty and lame but I'm like, well, I'm the adult, I guess this is what I'm supposed to do, I can't always be the teenager. You know?"

BK ROT feels like a routine, but the responsibilities of each worker change as time goes on. When M. cut down her processing time to a trim two hours, Renée and Sandy looked around for something else for her to do, and tasked her with trying to beef up BK ROT's social media presence. As the only media-savvy adult in the mix (and honestly—not very), I was asked to work with her. Figuring out how this works has been difficult—she didn't know how to write the first Facebook post, and I couldn't figure out how to explain the process to her, so I found myself dictating something and telling Sandy and Renée that we'd written it together. But the week after, I learned that Instagram was more her language, and she and I collaborated on trying to design an aesthetic for the Instagram account. Lots of emojis.

The addition of Domingo in spring 2016 changed the dynamics of labor at BK ROT. Domingo had experience running compost systems, and knew how to keep a pile active and happy. He also has a deep wellspring of charm that makes it easy for him to ask people to do things without offending them. A natural manager. As much as he likes BK ROT, he wasn't trying to be here forever, and he explained to me that he just wants to get the other youth workers to become more knowledgeable, more invested, and able to run the system themselves as smoothly as he does it. He was often frustrated in these efforts, but he handled the frustration well. He'd find a tumbler full of food scraps without browns, or a whole winter squash in a bin, and say "Ohhhh, someone's going to get a talking to," laughing with wide eyes and mock anger.

Until BK ROT had nonprofit status, the payment process was generally informal. Usually, at the start of the workday, Sandy and Renée would check in with each other to see if either of them remembered to pick up cash. One of them usually has to run to the ATM to get money from the BK ROT account.

Putting value on the labor that goes into composting is difficult when the primary models for dealing with food scraps are either volunteer labor or municipal collection, both of which appear free to people disposing of waste. Robin Nagle writes that the invisibility of the city's sanitation workers makes it easy to devalue sanitation work (2013). Ceci explains how the taken-for-grantedness of waste labor elsewhere in the city's compost system shapes users' attitudes toward BK ROT's food scraps drop-off: "Either not donating or not saying thank you are these different aspects that are a little bit concerning to me, . . . and I understand that this is how many other drop-offs work throughout New York City. But BK rot is trying to prove that . . . you bringing your food scraps in itself . . . this is still you disposing of your waste . . . in order to do something with it requires a lot of labor."

For Sandy, a central purpose of BK ROT is creating a framework for valuing the labor that goes into waste management:

> There's a lot of work in ecological practices that should be happening in the community, [it takes] a lot of labor that's not valued, that's generally seen as gardeners, or people who have time coming in and doing that for others for free, which kind of refers back to a level of privilege, and so this project mostly just tries to find ways to put value on that work and then create opportunities for the people who don't get to get that kind of work or the people who don't populate mainstream environmental gigs. And in this case it's Black, Latino, and immigrant youth. It's not that environmentalism isn't rooted in Black, Brown and immigrant communities, it's that those who typically benefit from eco-solutions through jobs are not them.

Sandy and Renée have both encountered community gardeners and community composters expressing judgment about BK ROT and other urban agriculture projects that charge for their services and products. For many in the urban agriculture world, volunteering and sharing resources freely are antidotes to capitalism, and allow people to live at least part of their lives outside of market forces. However, Renée points out the irony of an anticapitalist ethos in the hands of people who benefit from economic inequality: glorifying volunteering tends to privilege people with leisure time, spare income, and access to volunteer opportunities (Ghose and Pettygrove 2014).

BK ROT organizers have a vision of waste labor that contrasts with both the taken-for-granted labor of municipal collection and the virtuous volunteering of community compost. They demand that we treat waste laborers as experts, give them ownership over their work, and pay them fairly. They found that the most straightforward way to accomplish all this was by asking people to pay for BK ROT's services.

It became increasingly pressing for BK ROT to fight for this vision of waste labor as the city's curbside organics collection promised to expand into Bedford-Stuyvesant and Bushwick. DSNY collection was not ideal to BK ROT organizers—it moved the resources further than they needed to go, it failed to nurture local economies and local space, and it maintained city control of waste collection. However, it was also free, which made it hard for BK ROT to compete with. In 2016, in an informal survey of five BK ROT customers, I didn't find any who planned to continue using the service once the city began collecting residential organics in Bushwick. It's in part because of this that BK ROT switched entirely to commercial collection in 2017. The switch to commercial waste also meant fewer trips, more input, more income per pickup, and more efficient compost hauling for BK ROT bikers. The expansion of the NYC Organics Collection pilots halted in May 2018, before they began collecting in BK ROT's service area.

For projects with a social mission, navigating the green economy can threaten that mission. In her study of "green for all" projects that aim to create inclusive versions of sustainability, Alison Alkon (2012) describes how a farmers market that was meant to address food access needs in the predominantly Black neighborhood of West Oakland ended up mostly serving a white audience. She argues that, while laudable, the effort to plug social justice projects into the green economy can end up commodifying social justice struggles. In the West Oakland Farmers Market, racial justice effectively became something you could buy—if you had a taste for it, and if the price was right. In the green economy model of sustainability transitions, sustainability and social justice will prevail based on consumer choice to support sustainable, just operations. In their residential services, BK ROT has a similar relationship with clients, who are overwhelmingly white and middle-class. Customers have been able to purchase the experience of feeling good about their trash.

BK ROT is operating within the green economy, but it's also actively engaged in changing the rules of the green economy. By experimenting with waste infrastructure, and building a system that embodies their values, they've demonstrated alternative possibilities for urban composting. And through advocacy, they've helped to create regulatory and funding space where they can operate according to these values. While green economic models focus on the growth of green businesses, BK ROT's focus on workers means that the primary aim of their experiment has always been to create more hours for workers, create more jobs, raise wages, or make work better. In board meetings, as directors consider new purchases and new policies, labor is always a central concern: What will this do to the dynamics of work? Will it create more hours? Will it make the work easier?

Commercial hauling allowed BK ROT to make more income off of collection, but "earned income"—income from goods and services offered—has never been enough to adequately compensate the labor of BK ROT's workers. Instead, the majority of BK ROT's income has been from grants. This is frustrating in some ways—the grant-writing work of board members is never-ending, and it often winds up falling on the executive directors' shoulders—but it has in some ways offered the project a more flexible relationship with the forces of neoliberal waste management. Of course, grant funding comes with other commitments—to corporate, government, and foundation funders. But funders have been overwhelmingly supportive of BK ROT's vision, and grant funding has allowed the project to operate in the emerging market for compost services without having to compromise its values to make ends meet. While the BK ROT board dreams that the project might someday sustain itself off of earned income, for now we've accepted that the market for waste management simply isn't set up in a way that fees for waste services can adequately compensate the value of waste labor.

CONCLUSION: COMPOSTING UTOPIA

BK ROT works alongside the city's developing infrastructures for compost to challenge the politics of neoliberal waste management. Pushing back on the frameworks that define municipal collection (city-managed, export-centered, apparently free to residents) and the Compost Project (volunteer-run, free to residents), BK ROT has crafted a system in which waste labor deserves respect and compensation, and waste infrastructure is designed to be lived with. The project uses waste to challenge green gentrification by creating spaces that are strategically open to Bushwick's working-class, Latinx, and Black residents. Organizers' efforts to advance an alternative politics of waste and space have involved an ongoing negotiation with workers, neighbors, customers, regulations, microbes, aesthetics, funders, and the physical infrastructures of the city. Through these interactions, they have formed supportive relationships that anchor the project in the urban landscape.

IN RESPONDING TO the COVID-19 pandemic, BK ROT organizers needed to adapt this vision of just sustainability to a changing context with new risks and new economic pressures.

BK ROT suspended all services in late March 2020, when COVID cases started rising and the city went into lockdown. Workers were still paid for their normal hours while operations were on pause. Organizers tried to offer alternative composting options: days after suspending services, they posted a video to YouTube in which Ceci explained how to compost at home with a worm bin.[17] Staff pulled together worm bin kits that customers could order for bike delivery within BK ROT's service area.

Worker safety was a top priority, and Ceci worked with staff to develop safety protocols for when collection started up again. Collecting food scraps is a contact-free activity, and the US Composting Council determined that the risk of transmitting COVID through food waste was minimal (Millner 2020). Workers wore masks on-site, even though they were outside and risk of outdoor transmission was very low. The board of directors pooled funds to purchase herbal wellness kits for workers from a Brooklyn-based herbalist.[18] After a week's pause, BK ROT restarted commercial collection, but with fewer accounts because many restaurants had closed. After two weeks, they resumed Rabbit Roots, the residential collection service started by S. and R. With the closure of food scraps drop-off sites, residential collection was in high demand. BK ROT bikers also supported local mutual aid projects, delivering food and composting waste from food distributions.

It was more difficult to restart food scraps drop-off, which had been designed to facilitate interaction between youth workers and people bringing food scraps. Drop-off was also complicated because the two workers who ran it had previously used public transportation to get there, which carried significant exposure risk. BK ROT did not restart drop-off until July, with a modified approach: there was less face-to-face interaction, and users no longer helped process their own scraps. They found volunteers who drove workers to and from shifts. They kept paying workers their average

17 A worm bin is a system that uses a high concentration of worms to decompose organic matter fairly quickly into high-quality worm droppings. They can be relatively cheap; the BK ROT worm bins were made out of a plastic storage container with air holes punched into it. You can order the right kind of worm—red wigglers—online, and create a hospitable environment for them with some soil, food scraps, and shredded newspaper. Worm bins are great for apartments, because they move material quickly and have minimal smell (if you maintain them properly).

18 The kits included immune-boosting supplements and salve for sanitizer-chapped hands.

pre-pandemic weekly wages through the first months of the pandemic, until they found a way to safely resume services. This way, they didn't pressure workers to put themselves at risk for a paycheck.

For BK ROT organizers, facing COVID-19 meant squaring their vision of just sustainability with the changing landscape of waste management and risk management in New York City. Ceci describes how the project's values helped BK ROT navigate the shock of the pandemic: "[We were] trying to overcome operating out of a place of fear to operating out of a place that was rooted into our values. I think this was a lot for me in the initial arrival of the pandemic, and then . . . I think we pretty quickly regrounded in our values. It made it so clear, the importance to continue our work and continue modeling alternatives." Working from the project's values allowed BK ROT organizers to understand the day-to-day labor of hauling, composting, and coordinating as part of the long-term work of building just sustainability. This relationship between everyday actions and future building is what Ann Cvetkovich describes as "the utopia of ordinary habit" (2012, 191). Cvetkovich describes the importance of habit for utopia: "Habit can be a mechanism for building new ways of being in the world" (191). Everyday habits mean something else in utopia.

"Rooting into our values" demonstrates something special about the future orientation that I've characterized as composting. Composting responds to the present context but understands that context as transformable. Operating based on values—rather than on perceived constraints—allowed the BK ROT team to continue their ordinary, utopian work even as they faced the pressures of neoliberal waste management or the new challenges of COVID-19.

I've contrasted composting with a "landfill" orientation to the future, in which idealized futures are plopped down onto the imperfect present. Landfilling imagines new arrangements of people and environments, and these new arrangements often resonate with many people's values around sustainability and even justice. But in coming up with a solution to a bounded environmental problem, landfill utopianism tends to take the larger context for granted. In the curbside organics recycling pilots, DSNY tried to build a system that worked within a limited understanding of organic waste: as an expensive thing to haul, a source of greenhouse gases, and a potential resource for branding New York City as sustainable. Narrowing their focus allowed DSNY to build a system that was practical given existing conditions, but it meant that they failed to envision real transformation in many problems that connect with the city's waste system, such as environmental injustice, gentrification, and a lack of job opportunities.

The composting metaphor has been popularized by material feminist academics such as Donna Haraway (2016) and Jennifer Mae Hamilton and Astrida Neimanis (2018). But the most usable articulation I've found comes from Dior St. Hillaire, who became co-director of BK ROT in 2021. Before coming to BK ROT, St. Hillaire founded GreenFeen OrganiX, a worker-owned hauling and composting cooperative in the South Bronx. St. Hillaire is a Black woman who was born and raised in the Bronx, whose approach to sustainability is rooted in a critique of racial capitalism. St. Hillaire describes the work of GreenFeen OrganiX as "composting capitalism": "By composting capitalism, you're not necessarily eradicating it but you're putting it in the ground and taking what we all know and value and letting it grow into something else, let it grow into what's needed, because we should not be exploiting the world the way we're exploiting the world for capital gains" (quoted in Abello 2017). As an organics hauling and composting business, GreenFeen OrganiX is part of the green economy, and makes use of the functional space created by neoliberal waste management. However, as a worker-owned cooperative with strong ties to its neighborhood, the project is also building what J. K. Gibson-Graham call a "community economy": their economic activities are shaped by an ongoing ethical conversation about the kind of community that is being built through the specific ways they're organizing labor, exchange, and decision making (2006). St. Hillaire explains, "We take 'the best of capitalism' and . . . we break it down so much so that it's unrecognizable, and we create something that is for the benefit of all" (2020). Capitalism is their context, but they're building toward something else.

BK ROT organizers also approach their context as composters; they are building a system that consciously addresses present injustices as they work toward something else. Instead of simply designing a system that doesn't burden Black and Latinx neighborhoods with waste infrastructure, they're modeling a way that better waste infrastructure could be a benefit to overburdened neighborhoods. Instead of simply building a garden that's technically available for everyone, they're designing a space specifically for the Latinx and Black people who have historically lived there. And instead of simply eschewing capitalist exchange, they're trying to create a system that responds to the dehumanization and exclusivity of neoliberal waste management by hiring Latinx and Black youth and young people, valuing their labor, and developing their leadership. BK ROT organizers have never envisioned a system that existed free from inequality; they've built a system that tries to address present inequalities as they envision a world where those inequalities are not inevitable. For sustainability transitions policy and research, this demonstrates the necessity of a holistic account of the contexts in which transitions take place: imagining transitions in purely sociotechnical terms ignores the dependencies between infrastructures and inequality.

For BK ROT organizers, composting their context also means looking to the past to find alternative genealogies of sustainability, rooted in Black and Latinx identities and histories. On social media, organizers describe composting as an "ancestral practice," and remind followers that compost isn't some white innovation that embodies abstract, moralistic environmental values (BK ROT 2021). Ceci relates this historiography to the larger project of challenging waste systems: "Composting allows us to create our own models. It is a living practice where we get to reject dominant systems imposed on us and instead model how we want to live in relationship with each other and with the land. Composting is our ancestral legacy. It is a way in which our communities remember, take back power and create the world in which we want to live." Domingo likewise emphasizes that composting is not a new practice but an alienated one: "It's just been stolen from our history, and we don't know about it." By reclaiming the historical narrative—not just the legacy of environmental injustice but also the legacy of Black and Indigenous people's sustainable practices (Penniman 2018)—BK ROT uses the past as a resource with which to build toward new futures (Muñoz 2009).

Composting is an approach to utopia that accepts the complexity of the present while also assuming that fundamental transformation is possible. It creates practical utopias—utopias that are reachable from where we stand—without accepting already-existing definitions of what is practical. BK ROT composts their context, including the legacies of environmental injustice, racism, and economic inequality, and the legacies of Latinx and Black environmentalisms. The result is an experimental infrastructure for organics recycling that responds differently to environmental injustice, racism, and capitalism, and articulates a new narrative for how people can live with waste.

CONCLUSION

Lessons from the Compost Heap

> And honestly, I wish that what we were doing with the sector of waste at BK ROT was something that was possible to do with every sector. But the reality is that composting is a very simple, accessible technology that lends itself to this type of transformation.
>
> —Ceci, BK ROT

Composting can happen with minimal equipment and labor, locally, without much training. The experimental infrastructures for composting that I've described in this book are able to do this while also meeting other social justice and community-building goals, in part because the technology is accessible and flexible. Infrastructures for sustainability are not always so amenable: technologies such as solar panels, wind turbines, recycling facilities, and organic farms often require large spaces, expensive materials, international markets, and technical expertise, and are hard to implement otherwise.

Given these differences, what can we learn from community composters and microhaulers about how to address other problems around waste, sustainability, and social justice? What can we learn about sustainability transitions from the example of compost in New York City?

First, community-centered enterprises work in tension with neoliberal waste management. Many writers and activists have argued that market creation is an inadequate or misguided response to sustainability problems (Järvensivu et al. 2019; MacGregor, Ramasar, and Nicholas 2017; MacBride 2011). In New York City's curbside organics program, market forces have favored large-scale waste management approaches over localist approaches, or approaches aimed at creating good compost. Composting has been treated as an individual service, rather than a municipal good. However, community composters and microhaulers have found ways to resist or co-opt market forces or change the rules of the market. Building community economies around labor, rather than waste flows, has been one strategy for this.

Second, community composting demonstrates the relevance of experimental infrastructures. Studies of sustainability transitions have pointed out problems both with top-down interventions that impose a single definition of sustainability onto a diverse landscape (Dillon and Sze 2018; A. Smith 2012), and with "small-steps" approaches that assume sustainable worlds can be built through the accretion of individual actions (MacBride 2011; Maniates 2001). Community composters and microhaulers demonstrate a middle path between these approaches: *experimental infrastructures*. Experimental infrastructures allow people to inhabit their worlds in new ways, build new forms of community, and reshape the discourses of larger institutions.

Finally, the experimental infrastructures I describe in this book enable a specific experience of hope that is relevant for other sustainability efforts. Community composters in New York City build new possibilities out of their local contexts, creating alternative systems that function here and now. I've described this kind of world-building as *composting utopia*. Composting shifts the emotional core of sustainability transitions from specific sustainable futures to alternatives that are available in the present, a move that should offer inspiration for other transitions in the making.

SETTING THE RULES OF THE WASTE MARKET

In building an economy for organics recycling, participants have disagreed about the best way to assign value to discarded organic material and waste labor.

Nearly everyone involved in building compost systems for New York City has agreed that it's important to put a positive value on organics. Community composters would often correct me when I used the term *food waste* to describe the materials they received: *it's not waste; it's a resource*. The first goal of outreach for master composters is to "help New Yorkers recognize food scraps as valuable resources that can benefit their community" (DSNY 2015b, 6–3). Calling it a resource has legal implications for small-scale haulers, who have found more freedom hauling *resources* than hauling *waste*.

But treating waste as inherently valuable creates problems for the people working with it. For instance, organizers at Compost for Brooklyn almost always find a bag of rotting food scraps at the gate when they show up to open up the space. Nancy explains that the people who leave these bags think they're composting but they're not—they're just leaving garbage at the garden. Some days she just throws the bags away in frustration. Facing the same problem at their own gates, BK ROT organizers posted an image to their Instagram illustrating the difference between "compost" (a small mound of dark brown material) and "food scraps" (a multicolored jumble of vegetable scraps and eggshells). The caption reads:

> Just so we are all on the same page.
> Compost is what we offer to the community (compost ≠ dirt)
> Food scraps is what we collect from the community on Sundays from 12–3pm. (August 7, 2021)

The distinction between waste and resource is an important one for community composters. In a comic about BK ROT, smell lines and gross-out faces mark food scraps as currently disgusting, despite their potential value.

In overvaluing waste, this new economy has the potential to undervalue waste labor. At BK ROT, Sandy and Renée perceive the free composting provided by the Compost Project as a threat to their fee-based services, which are meant to compensate the labor of waste management. Projects like BK ROT challenge the development of a market in which waste is valuable but waste labor is not.

An alternative framing treats composting as a social good—putting waste to beneficial reuse. In contracts with haulers and processers, at both large and small scales, discarded organic materials are treated as waste. Hauling fees and tipping fees compensate the labor involved in waste management, but markets for waste services can also get messy. The competitive market for waste hauling has previously been responsible for concentrating New York City's waste infrastructures in North Brooklyn and the South Bronx, and for creating the overlapping diesel truck routes that pollute these neighborhoods (Wood 2019; Casey and Mehrota 2011). These markets can often create exploitive working conditions for the people doing the actual hauling (Reno 2016).

At Peninsula Compost Facility in Wilmington, Delaware, tipping fees drove a processing disaster. Designed and operated to maximize capacity, Peninsula accepted organic waste from New York City's curbside organics program and made minimal efforts to decontaminate this stream. The large volumes of poorly sorted material made the composting process inefficient, generating noxious odors that spread for miles.

Disasters like Peninsula are common in waste service markets, but individual firms and players sometimes find ways to act otherwise. McEnroe Farm, which also received a significant fraction of the organics collected in the curbside pilots, accepts only as much organic waste as it can process, and is selective about what it takes in. The majority of compost produced by McEnroe is used on the farm as a soil amendment, and the rest is sold at hardware stores and food co-ops. Its tipping fee is low compared to other composters, in part because it can cover costs through compost sales. McEnroe's facility has remained small since it began operating in the early 1990s. This is partly because its capacity for inputs has tracked with the needs of farmers and the dynamics of the market for finished compost.

The new market for organics recycling services has also created a space in which microhaulers have found residential and commercial customers who are interested in reducing their ecological footprints and doing something positive with their waste. This was particularly notable during the pandemic, when several microhaulers stepped up to handle residential waste that was no longer being picked up by the city (Nierenberg 2020). Some of these microhaulers, such as BK ROT and GreenFeen OrganiX, use the organics recycling market to pursue social justice goals—a move that Dior St. Hillaire describes as "composting capitalism." St. Hillaire describes how the project uses the market to create something that is not necessarily subject to market rules: "It feels like family when you go up in there. It doesn't feel like capitalism" (Abello 2017).

The emerging market can support multiple models of organics recycling, but it's not an even playing field. McEnroe's limited capacity and its need for high-quality inputs make it hard for it to expand its presence in the waste management market. And pressures on microhaulers—such as licensing, rate caps, and a lack of processing space—make it difficult for them to function at a profit. Industrial compost systems designed for waste management, in contrast, are flexible in their capacity and able to achieve economies of scale. The rules that have been set for the organics recycling market privilege managing waste over creating compost, consumer convenience over well-sorted inputs, and large-scale systems over local ones.

Some organizations have gone beyond working within the market, and have found ways to challenge it. While BIC set a cap on the rate that microhaulers could charge by weight or volume, BK ROT developed an alternative pricing model based on the amount of labor required, reframing the waste market around labor rather than materials. They have also worked with Vokashi, Common Ground Compost, and Green Feen OrganiX—the microhaulers' alliance—to successfully appeal for an increased rate cap, a more expansive definition of microhauling, and special protections for microhaulers. Challenging market rules requires serious work on the part of activists to reframe the logics of waste systems, and a willingness to wait in precarious informality while articulating their own alternative frameworks.

EXPERIMENTAL INFRASTRUCTURES IN SUSTAINABILITY TRANSITIONS

Studies of sustainability transitions have pointed out insufficiencies with both top-down and bottom-up approaches to building sustainable worlds.

In the top-down narrative, new technologies and policies can reorganize a wasteful society into something more sustainable: changing the world requires large-scale interventions, whether directly creating new infrastructures that can support sustainable systems, or by using market forces to encourage individuals and firms to pursue a centrally defined version of sustainability. In practice, top-down interventions are not always able to achieve the desired results. One reason for this is a mismatch between top-down interventions and the daily lives of people affected by them. Adrian Smith argues that top-down approaches treat civil society as a "problematic medium" that needs to be managed in order to create sustainable systems (2012). In New York City, where curbside organics collection managed to divert only 11% of the compostable material in the areas that it served, top-down transition ran into serious obstacles.

In bottom-up approaches, individual behaviors and small, local interventions will add up to larger change, and inspire people to make a difference. Every banana peel diverted is a small step toward a sustainable world. Small local enterprises will eventually become larger and saturate the market with sustainable transactions. Changing the world just requires small-scale consumer and entrepreneurial action in order to generate larger-scale changes in industry. Samantha MacBride suggests that this approach is prone to busy-ness, generating lots of activity but little measurable progress toward sustainability. MacBride argues that the small-steps approach relies on consumer preferences, individual decisions, and the goodwill of CEOs, none of which is a reliable resource in the project of crafting sustainable systems (2011).

In writing about experimental infrastructures, I'm offering another account of the construction of sustainable futures. In experimental infrastructures, people co-construct systems and communities that can work toward locally defined ideas about sustainability. Rather than tackling sustainability problems as global abstractions or as issues of individual morality, people build new sustainable systems in relationship with place and with one another. These interventions are not idealistic escapes from reality; they remake the systems they inhabit.

Experimental infrastructures are often informal, operating in the cracks to respond to needs that are not met by formal systems. This makes them more precarious than formal systems, but it does not mean that they're determined by their conditions. The designers of experimental infrastructures do more than just "make do" with what they have; they challenge narratives of what they have and what they can make of it.

Around the United States, experimental infrastructures for just sustainability have gone beyond survival to demonstrate alternative approaches that are available in the present. In Georgia, the Sapelo Island Cultural and Revitalization Society is repurposing the ancestral knowledge of the Saltwater Geechee people to grow traditional crops on commonly owned land (Heynen 2020). In Mississippi, Cooperation Jackson offers a model of "sustainable community development, economic democracy, and community ownership" that includes organic farms, community compost, and local production (Madeson 2022). Soul Fire Farm in upstate New York demonstrates how food systems could be designed to challenge food apartheid, disenfranchisement of farmers of color, and eco-colonialism (Penniman 2018). Community solar projects in Vermont prefigure the possibility of energy democracy (Stephens 2019). Experimental infrastructures create space in which people can practice new ways of connecting with one another and understanding their worlds. They open up new possibilities.

The term *experimental infrastructures* came out of grant-writing work with BK ROT. The term offers a kind of futurity that is appealing to funders: trying something out at a small scale in order to later try it at a large scale. This is conversant with a popular narrative of sustainability transitions: the multilevel perspective depends on experiments in protected niches that, if successful, reshape sociotechnical regimes (Geels 2005, 2011). Research on "urban living labs" and "bounded socio-technical experiments" examines the role of formal experiments conducted by research institutions and city planners (von Wirth 2019; Bulkeley et al. 2016). Harriet Bulkeley and collaborators describe systems that are "co-created by research organisations, public institutions, the private sector and community actors," as a means to develop new knowledge that is "explicitly captured and used to inform the process of creating urban sustainability" (2016, 13). These accounts recognize the multiple ways in which experiments can reshape the social world, but generally take an empiricist approach to experiments: they are a method of testing new ideas.

While I share this hope that community composters will be able to prove the effectiveness of their own approaches, I've been imagining the term in conversation with STS writing on experiments. In STS, there is a long tradition of examining experiments as a performative mode of knowledge production and as inflection points for reshaping the social world (Shapin and Schaffer 1985; Latour 1993; Galison 1987). Hans-Jorg Rheinberger waxes utopian when he describes experiments as "machines for producing the future." A good experimental system is one that can "give answers to questions that we are not yet able to formulate clearly" (2012). Experiments don't produce knowledge; they produce questions along with communities who can answer them.

In the same way, experimental infrastructures are doing more than producing knowledge or testing things out: they're producing new ways of thinking about sustainability and new communities who can work toward new goals. Per Casper Bruun Jensen and Atsuro Morita, all infrastructures are "ontological experiments" that "bring to life novel relations and objects that do not correspond with the intentions of policy makers and engineers" (2015, 83). Experimental infrastructures belie the standard account of sustainability transitions, in which a transition is oriented toward the end goal of a sustainable system. Instead, they reveal emergent possibilities for human relationships with our environments and with one another, new and surprising kinds of identity, labor, and community that are imaginable in the pursuit of earthly survival.

Community composters aren't trying to figure out if community compost will or won't work as a solution to New York City's waste problems, they're building worlds in which community compost *does* work, and inviting people to stop by and grab a pitchfork.

In the following sections, I describe three ways that community composters have an impact beyond their immediate scope: they allow people to relate to waste in new ways; they assemble communities around new visions; and they form generative partnerships with existing institutions. Experimental infrastructures invite us to examine sustainability transitions as more complex, emergent phenomena, in which grassroots innovators reorganize their worlds.

ENABLING NEW RELATIONSHIPS WITH WASTE

Community compost allows people to relate to waste in new ways. On-site, volunteers get involved with the work of waste management: they pick up shovels, chop up garbage, and squish their hands in the compost. At home, users practice new habits, such as sorting out food scraps, cleaning out compost buckets, and dragging food scraps to the farmers market. It can even make composting present for people who aren't involved: compost buckets on front stoops, the smell of food scraps when people walk by the farmers market, neat rows of bins at Grand Army Plaza, and strangers carrying bags of garbage down the street.

This is one of the reasons why community composting appeals to DSNY: it trains people in how to engage with a compost system. Advocates of both community and industrial composting have recognized community compost as a tool for building a "culture of composting" (Paull 2019). The Community Composting Report explains that the Compost Project and community compost sites spread the good word about organics recycling: "They raise awareness about what compost is and its benefits through outreach and education, and by using compost to grow food and care for green spaces in NYC neighborhoods. Making and using compost locally demonstrates firsthand to New Yorkers that apple cores and eggshells are not garbage—they are useful resources" (DSNY 2014, 6).

These cultural shifts are important, but discard studies has challenged the idea that raising awareness among waste generators can meaningfully address waste problems. Ideally, awareness will make consumers want to "do something" about waste, but the actions that are available to consumers—reducing, reusing, recycling—are geared more toward busy-ness than impact. Individual interventions will be ineffective if powerful actors don't simultaneously pursue systemic and infrastructural changes (Schouten, Martin, and Tillotson 2015; MacBride 2011). Awareness campaigns work within the limits of neoliberal waste management, and emphasize individual responsibility over systemic change (Maniates 2001).

Experimental infrastructures for composting can create space for a more open-ended form of ecological consciousness. Community composting can cultivate "response-ability" toward food scraps, polluted soil, and community (Haraway 2008; Schrader 2010; Kenney 2019). Drawing on Donna Haraway and Astrid Schrader, Martha Kenney explains that "the feminist ethic of response-ability focuses not on being responsible but on learning how to respond and 'opening up possibilities for different kinds of responses'" (2019). Experimental infrastructures for composting allow people to practice responding differently to urban wastes. At C4B, responding to food scraps is tied up with responding to neighbors: watching kids, inviting people in, chatting up passersby. Projects like BK ROT or Green Feen OrganiX demonstrate that it's possible to respond to food waste while at the same time responding to environmental injustice and gentrification.

Experimental infrastructures also allow us to practice new emotional responses. People might respond to waste with feelings of connection over shared work, or hope for better worlds, but our responses often include disgust. We meet sticky stenches that cling to our nose hairs; half-recognizable leftovers squelching underfoot; worms, centipedes, soldier fly larvae, and other less familiar denizens of decay. We respond with gross-out sounds and scrunched up noses, but compost piles also give us space to feel disgusted without running away.

Through repeated interactions, new responses to waste transform into habits, and develop their own momentum. New habits have the potential to embed future potentialities in present practices. Ann Cvetkovich describes this as "the utopia of ordinary habit," in which habits become "mechanism[s] for building new ways of being in the world" (2012, 191). Gay Hawkins argues that waste habits can establish new ways of understanding the world: "In the emergence of new waste habits, an experiment with another social imaginary, whether it's explicitly identified or not, lurks in the background" (2006, 14).

As responses become habits, response-ability might morph into responsibility. Community composters have put themselves in a position where they are responsible for managing their neighbors' garbage, and New Yorkers are not planning to stop throwing things away anytime soon. In my weaker moments, I've regretted this commitment. There were many mornings where the last thing I wanted to do was get out of bed and face the city's waste stream. A fellow volunteer likens our weekly chore of turning the compost pile to her experience of church: while turning the compost is "a horrible ritual we're forced to take part in," she says that "the sort of consistency that it provides every week . . . kind of keeps people engaged with it." Experimental infrastructures are both grassroots innovation and ongoing commitment.

When I asked Ceci how BK ROT is changing the landscape of waste in New York City, they explained that they're "modeling a way in which it is possible to restructure our relationships to waste." Experimental infrastructures for composting can raise awareness, but they also offer an opportunity for users to develop their own relationships with waste, and practice new ways of responding to waste problems. Response-ability, responsibility, and habits don't immediately accomplish the kind of systemic change that is necessary to actually address the ongoing harm being done by waste systems. But they go beyond the project of training people into a "culture of composting." Instead, they encourage people to develop personal, meaningful relationships with waste, soil, neighbors, and urban spaces. This is fertile ground in which to grow new systems.

ASSEMBLING NEW COMMUNITIES OF PRACTICE AND CONCERN

"Community composting" sounds like a situation where communities create solutions to shared waste problems. But J. K. Gibson-Graham point out that community is not a pregiven entity: it's something that emerges in interaction (2006). The "communities" of community composting are assembled and structured through strategic efforts of inclusion, exclusion, and organizing.

At BK ROT, the conscious construction of a community is an important and ongoing task. Organizers have tried to make sure that BK ROT supporters aren't just the white newcomers endemic to Bushwick garden spaces but instead a crowd that looks more like the rest of the neighborhood—Black and Latinx and working-class people. This has involved making their space inviting and open rather than claimed and personal, filling it with Latinx-produced art, and knocking on doors to check in with neighbors about project design.

The construction of a BK ROT community has also involved careful networking with similar projects and political allies. Sandy and Renée have built connections with other projects that center people of color in sustainability work, such as Brotherhood Sister Sol in Harlem and Loop Closing Compost in Washington, DC. They are more careful about connecting with groups that are interested in the technical innovations of BK ROT (bike-based compost hauling) without the social innovations (good green jobs for youth of color), but have still worked with other microhaulers in order to push for better policies. In their efforts to build a community, BK ROT organizers have been careful to communicate the values of the project first and foremost. BK ROT does not simply demonstrate that bike-based composting can divert waste, but also shows that bike-based composting can create good green jobs and accessible green space.

Organizers have also worked hard to bring people and organizations with power and money into the BK ROT community. By making the project appealing to city council members, to people at DSNY, and to funders, they have been able to move their discourses into public conversations and attract useful supporters.

The Master Composter Manual offers instructions for bringing communities together. It explains how to organize a management team, attract volunteers, find supporters, and communicate with neighbors. Different compost sites do this work in different ways. C4B functions as a hub of activity, in part thanks to Nancy's outgoing personality: organizers greet people on the street and offer childcare to neighbors. At Earth Matter, organizers circle up volunteers at the end of the workday to name a new windrow, giving people space to reflect on the work they've done together. Big Reuse assembles a more abstract community of users, who don't necessarily communicate with one another, but nonetheless share practices and values in their relationship with community composting.

Experimental infrastructures are also reshaped by their communities. Writing about sustainable living experiments, Noortje Marres suggests a two-way power dynamic in public experiments: they "provide important sites for a politics of contestation, and not just for a politics of legitimacy or 'enrolment'" (2009).

Sandy and Renée initially conceived of BK ROT as a system that keeps waste local, runs on sustainable energy, and values the youth of color they work with. Putting all these pieces into practice has required constant physical, microbiological, interpersonal, and emotional negotiation with the realities of the project. They tinkered with the design of bins to keep rats out and allay neighbors' hygiene concerns. They tested different bags and buckets for collecting waste. They altered the balance of nutrients in their compost piles and added air and water. They tried out possible alliances with restaurants and haulers. They transitioned from residential to commercial hauling in the face of the city's expanding curbside organics collection, and then back again during the COVID-19 pandemic. Sandy and Renée learned how to take on the various social roles needed to work with youth, and what each role can do: sister, aunt, boss, mentor, friend. Through all of this tinkering, they've navigated differences between their own ideas about compost, green space, and environmental justice, and the ideas of their workers, neighbors, and supporters. This tinkering has grounded the hopeful visions of BK ROT in the historical realities of Bushwick, and an ongoing struggle for survival.

BUILDING IN PARTNERSHIP WITH MUNICIPAL INSTITUTIONS

DSNY and the Parks Department have recognized an overlap of interests with grassroots composters: cultivating awareness of compost, managing park waste, and caring for green space. Institutions have offered material and institutional support, and have played a role in shaping the landscape of grassroots composting in New York City. In the process, grassroots composters have shifted the ways that municipal institutions approach organics recycling.

Ananya Roy characterizes formal and informal development as interdependent modes of city making, with each approach shaping the conditions in which the other operates (2009). Municipal institutions have always shaped grassroots composting: many community compost sites got their start as projects in the Master Composter Certification Course, operated by DSNY and the Parks Department, which used to require trainees to create compost systems in their communities. Even the projects that weren't run by master composters—such as the Lower East Side Ecology Center (LESEC) or the Fort Green Compost Project (now Earth Matter Compost)—formed in response to the lack of organics recycling services offered by the city.

But grassroots composters have experimented with new approaches to organics recycling that went beyond the Master Composter Certification Course. Greenmarket drop-off and commuter composting, developed by the LESEC and the Western Queens Compost Initiative (now part of Big Reuse), respectively, offered alternative solutions to the collection problem that fit easily into people's weekly lives. Microhaulers developed new business models for local composting, using bikes or personal vehicles to haul food scraps, and composting in community gardens. The new approaches to composting have often centered local values: a better relationship with waste, stronger community ties, and job opportunities for young people.

Informal development will often meet the needs of formal planners. DSNY recognized that community composters could meet their goal of cultivating a "culture of composting": they demonstrated what compost is, educated people about why it was important, and taught people how to separate out food scraps. Compost Project representatives approached a handful of community composters to propose partnerships: LESEC, Big Reuse, Earth Matter, and Added Value. Formalized into NYC Compost Project host sites, these projects receive support in the form of salaries, equipment, feedstock, and space. Most community composters, and even some microhaulers, receive some form of municipal support, whether it's space and browns through the Parks Department, or training and advertisement through the Compost Project.

Municipal support has offered legitimacy and stability to community composters, but it comes with conditions. The rules of the partnerships between host sites and DSNY makes it unlawful for them to charge for collecting organic waste or for selling compost.[1] Pandemic-related funding cuts demonstrated that municipal support is not guaranteed, and some projects struggled when the city's priorities changed.

Informal systems can also change the conditions in which formal systems operate. Grassroots composters have been able to shape how municipal institutions approach compost. Greenmarket drop-off has received ongoing support through GrowNYC. And microhaulers worked with the Business Integrity Commission to secure more favorable regulations that allowed them to expand their operations. After municipal organics collection was suspended at the beginning of the COVID-19 pandemic, networks of community composters turned out to campaign for universal, mandatory organics collection—a proposal that was passed into law by city council in June 2023 (Rubinstein 2023). DSNY has even tentatively explored processing city-collected organics at midscale community composting sites.

The interactions between grassroots composters and municipal institutions demonstrate the dynamic relationships between experimental infrastructures and the state. The state shapes the conditions in which experimental infrastructures operate, occasionally supporting experiments that meet state needs, and setting limits on the kinds of experimentation that are possi-

1 As previously mentioned, monetizing composting carries its own dangers. I deal more directly with the tensions between "dirty money" and volunteerism in chapters 3 and 4.

ble. But the organizers of experimental infrastructures also have a platform to challenge the way that the state operates, put forward new ideas about how to organize the world, and mobilize support for alternatives.

COMPOST AND HOPE

"Compost intentions" are a prize in BK ROT's annual #BlackGold fundraiser. The website explains what they mean by this: "We will write your intention on compostable paper using natural paints and offer it in a bundle with herbs and woody materials. As this special batch moves through our windrows, we hope our collective prayers will reach our eco-community through our decomposer family and finished compost as it returns to the earth" (BK ROT 2020). I wasn't on the committee that came up with the compost intentions, but when Ceci told me about them, it spoke to a truth about composting that I knew deep down: something special is happening in the compost pile. Composters get a front-row seat to the drama of destruction and creation; it's a place where people can touch the universal principle of life. I've heard even the soberest, most scientific composters describe composting as "magic."

Compost culture holds reverence for a process that defies human mastery. Even with the discipline and expertise that composters learn, many recognize that the magic of composting is outside of their control. Compost scientist Sally Brown describes this relationship in a 2012 *BioCycle* editorial: "Decomposition of organics is a magical process. But the key to the magic is that it will happen if you just set the stage" (52).

The compost intentions demonstrate a relationship with hope that exists in experimental infrastructures: these projects animate an intention, but they're just "setting the stage" for the actual magic of social transformation, which is outside of their control. Instead of hoping for an escape from the present and into a better future, experimental infrastructures *compost their context* to build alternative worlds out of what we have now. Composting responds to the complexity of present conditions, while acknowledging that the present is just organic material, ready to be transformed.

I'm contrasting composting with another experience of hope in relation to sustainability transitions: the promise of specific sustainable futures that can bury the problematic past and present. I've characterized this as landfill utopianism, and it's present in both top-down and bottom-up transition narratives. The future-oriented imagery of mainstream environmentalism often involves new technologies that will save us: oceans of solar panels, electric vehicles everywhere, a countertop food waste digester in every home (Jesse Goldstein 2019). It might also be personal transformations that will save us: people will one day realize how important it is to recycle, or eat organic, or turn off the lights when they leave the room (Maniates 2001). It can feel exciting to imagine a future in which our environmental problems have been solved—it might even give you what you need to get out of bed in the morning. But these images are abstract utopias: they cultivate a relationship with hope that is counterproductive to the project of social change. It's hard to compost your context if you're not inhabiting it.

Lauren Berlant used the term *cruel optimism* to describe how "the American dream" and other accounts of the good life motivate people to invest their hopes and energies in systems that undermine the possibility of achieving those hopes (2011). Sustainable futures can become cruel when they limit the possibilities of present interventions. Techno-optimism legitimates unsustainable use of resources in the present, and encourages policymakers to delay environmental action in the hope that future technologies can solve any environmental problems we create with present actions (Gardezi and Arbuckle 2020; Allwood 2018; Asayama and Ishii 2017). Grassroots, community-driven solutions can also be cruel, in that they are often materially insufficient to address the problems they're supposed to address (MacBride 2011). Riding a bike and other personal lifestyle changes feel promising, but in the absence of systemic changes will fail to accumulate into the kinds of impact that are needed (Maniates 2001; King 1995).

Cruel optimism relies on a belief about how the future is going to unfold: things will get better, either through external changes that will transform our situation, or through hard work within the present system. As Berlant points out, cruel optimism allows people to survive in current systems but can subvert efforts to both recognize present conditions and imagine alternative futures.

Composting is hopeful, but it does not make claims about the future. Instead, hope is cultivated when we bring new intentions to the complex present.

The community composters and microhaulers that I've described in this book are composting food scraps, but they are also composting neoliberal waste management, gentrification, and environmental injustice. They treat these systems as context, but not constraint, and feel out alternative ways of inhabiting them. The unending waste stream of New York City becomes a resource that can nourish local ecosystems. Emerging markets for organics hauling become places to experiment with community economies. Composting doesn't promise that things will get better; it demonstrates that things could be otherwise.

While cruel optimism makes promises, composting is about process. People find hope through the everyday work of living differently with their environments. People witness personal transformations as they reexamine their relationships with their waste and with each other, and material transformations as waste becomes resource. The everyday kludge jobs of maintaining these systems make them more meaningful to the people within them and around them. Sometimes the work is fun—a hobby that allows you to connect with neighbors and play with worms. At other times, it's exhausting, poorly compensated, and prone to burnout. But it almost always feels like something different from waste-as-usual. The labor is part of concrete utopia, demonstrating that we can inhabit the world otherwise.

Before working with community composters, I had learned to recognize this feeling in the concrete utopias of punk houses and DIY scenes, where remaking the world is an everyday activity. I don't often encounter this feeling in mainstream sustainability spaces.

I've described composting as a property of grassroots projects, but I don't want to imply that composting is *always* or *only* available at the grassroots. Grassroots infrastructures often rely on composting because they need to make do with what they have. But many grassroots projects work without outside input, disconnected from context, or in pursuit of specific, inflexible ends. Grassroots organizers are no strangers to cruel optimism and other, less flexible, relationships with hope.

And grassroots projects are not the only places where composting happens. DSNY has also composted its context through the NYC Compost Project: they've meaningfully worked with and supported existing projects, trying to figure out how to draw the city's grassroots composters together into something new. While DSNY has occasionally limited what community composters could do, they've also created a platform that has supported hundreds of experimental infrastructures for organics recycling that respond to different needs in different neighborhoods: multiple utopian visions that enact local ideas about the sustainable city.

The city's optimism about organics recycling becomes cruel when it is constrained by the logic of neoliberal waste management. The curbside composting pilot that started in 2013 was an exciting promise. Press releases described "the final frontier" of recycling, promised zero waste to landfill by 2030, and projected the largest organics recycling system in the world (Office of the Mayor 2013b; DSNY 2015a, 2016). But it failed to connect with the calculus of city budgeting, with existing systems, or with the realities of everyday New Yorkers. Because it failed to deliver on the promise of economic savings, the program was unable to weather the budget cuts that came with the COVID-19 pandemic.

So what might it look like for a megacity to compost its context?

In March 2022, Sandy Nurse, newly appointed chair of the Sanitation Committee of NYC City Council, stood in front of Brooklyn Borough Hall, alongside the borough president, the city comptroller, and fellow city council members (Edwards 2022). She booed at the "Rat Lobby," a pair of humans in rat masks carrying signs that said "Cut DSNY Budget Save the Rats." As Eric Adams, the newly elected mayor, threatened to cut the DSNY budget—including organics collection—Sandy and others defended compost. Sandy argued that composting wasn't just a lofty environmental goal, it's an environmental justice solution: it means cleaner streets, less waste flowing through transfer facilities, and the opportunity for all New Yorkers to participate in the sustainable city.

This vision of organics recycling for New York City is built on local knowledge and local concerns. It composts what organizers know about New York City and New Yorkers. Instead of an imagined future in which organics collection can meet the needs of environmentally-minded residents, this account of compost responds to existing everyday concerns. It's built around values that are more widespread than abstract environmentalism: clean streets, clean air, and green spaces.

In June 2023, as I was finishing up edits on this book, city council passed a package of bills inspired by this vision. It includes mandatory composting for all of New York City by October 2024, and support for community composting. Sandy was one of three lead sponsors (Rubinstein 2023).

Industrial-scale metaphorical composting can take place when community organizers force industry and policymakers to build for a future that is grounded in present needs. It is not a given, but it is possible.

I've spent a lot of time in this book trying to explain why grassroots infrastructural experiments are good and useful in the face of large-scale problems. I don't want to deny very valid arguments about the co-optation of green-for-all initiatives, busy-ness, neoliberal retrenchment, or the valorization of volunteerism. All of these phenomena have been described in great detail by a host of prior scholars; I encourage you to read the excellent work of writers such as Nina Eliasoph, Samantha MacBride, Julie Guthman, and Alison Alkon.

I also don't want to ignore the necessity of large-scale changes in the very near future, to address our intertwined environmental and political crises. Alternative systems are needed, posthaste, in order to secure a livable world for present and future generations. Small-scale local interventions, on their own, and uncoordinated, are unlikely to meet the demands of this moment.

The lesson from the compost heap is that transition can start where we are. Responding to the complexity of the here and now can offer meaningful alternatives to taken-for-granted narratives of transition. J. K. Gibson-Graham argue the importance of reorienting around the present: "Yet if we must necessarily 'start where we are' to build ethical economies, what is the usefulness of simply judging such practices for their divergence from certain values? It would seem more positive and practical to treat the *existing* situation as a (problematic) resource for projects of *becoming*, a place from which to build something more desirable in the future" (2006, 98). The builders of experimental infrastructures start where they are, composting what they have in order to create something new.

When I step into a community compost site, another world feels possible. I become part of a system that makes a new kind of meaning out of waste. No longer a solitary weirdo with an inexplicable affinity for rotting food, I find solidarity with friends and strangers who have likewise fallen in love with decay. The bag of frozen food scraps that my roommates had complained about is now a useful input to an alternative food system. The hard angles of a concrete world give way to the organic curves of windrows and well-fed wildflowers. In spaces like these, it feels possible to hope for a new relationship with the urban ecosystem.

Composting utopia—building new systems that respond differently to the present—can expand people's imagination about sustainability and build new communities that are working toward alternative futures. Elsewhere, composting their local contexts and local problems, other organizers have crafted their own experimental infrastructures that offer situated visions of just transition. Projects such as the Sapelo Island Cultural and Revitalization Society, Cooperation Jackson, Soul Fire Farm, community solar projects in Vermont, and many others ask new questions about sustainable futures: Who is sustainability for? What is a good green job? Who designs our ecological futures? They create new worlds that challenge taken-for-granted narratives about sustainable futures, and nurture new communities that can respond differently to the present.

Dior St. Hillaire characterizes the sense of possibility in places like these: "It feels like everything we've been dreaming about, everything we've been hoping for, you go to Green Worker Cooperatives and you're like, it's possible, it's all possible. We just got to work together" (quoted in Abello 2017).

So let's do that. Let's build compost piles out of our imperfect present, and feel them get hot. Let's set the stage for something beautiful.

References

Abello, Oscar Perry. 2017. "Bronx Worker Cooperative Plans to 'Compost Capitalism.'" Next City (blog), August 9. https://nextcity.org/urbanist-news/bronx-worker- cooperative-compost-sustainability.

Abumrad, Jad, and Robert Krulwich. 2013. *Poop Train*. http://www.radiolab.org/story/poop-train/.

Agnotti, Tom. 2008. "Is New York's Sustainability Plan Sustainable?" Presented at the Association of Collegiate Schools of Planning and Association of European Schools of Planning, Chicago, IL, July. http://www.hunter.cuny.edu/ccpd/repository/files/is-nycs-sustainability-plan-sustainable.pdf.

Agyeman, Julian. 2008. "Toward a 'Just' Sustainability?" *Continuum* 22 (6): 751–56. https://doi.org/10.1080/10304310802452487.

Agyeman, Julian, Robert D. Bullard, and Bob Evans. 2002. "Exploring the Nexus: Bringing Together Sustainability, Environmental Justice and Equity." *Space and Polity* 6 (1): 77–90. https://doi.org/10.1080/13562570220137907.

Ahmed, Sara. 2004. *The Cultural Politics of Emotion*. Routledge.

Akese, Grace A., and Peter C. Little. 2018. "Electronic Waste and the Environmental Justice Challenge in Agbogbloshie." *Environmental Justice* 11 (2): 77–83. https://doi.org/10.1089/env.2017.0039.

Alaimo, Stacy. 2010. *Bodily Natures: Science, Environment, and the Material Self*. Indiana University Press.

Alexander, Catherine, and Joshua Reno. 2012. *Economies of Recycling: Global Transformations of Materials, Values and Social Relations*. Zed Books.

Alexander, Ron. 1996. "Innovations in Compost Marketing." *BioCycle*, October.

Alkon, Alison Hope. 2012. *Black, White, and Green: Farmers Markets, Race, and the Green Economy*. University of Georgia Press.

Allison, Hilary. 2015. *Compost Comix*. New York: self-published by the author.

Allwood, Julian M. 2018. "Unrealistic Techno-Optimism Is Holding Back Progress on Resource Efficiency." *Nature Materials* 17 (12): 1050–51. https://doi.org/10.1038/s41563-018-0229-8.

Amankwaa, Ebenezer Forkuo. 2014. "E-Waste Livelihoods, Environment and Health Risk: Unpacking the Connections in Ghana." *West African Journal of Applied Ecology* 22 (2): 1–15.

Amenta, Edwin. 2006. *When Movements Matter: The Townsend Plan and the Rise of Social Security*. Princeton University Press. https://doi.org/10.2307/j.ctv173f2v6.

Amiri, Farnoush. 2018. "NYC Pauses Expansion of Ambitious Residential Composting Program." *Gothamist* (blog), June 1. http://gothamist.com/2018/06/01/nyc_composting_sanitation.php.

Amoako, Clifford. 2018. "Emerging Grassroots Resilience and Flood Responses in Informal Settlements in Accra, Ghana." *GeoJournal* 83 (5): 949–65. https://doi.org/10.1007/s10708-017-9807-6.

Anguelovski, Isabelle. 2014. *Neighborhood as Refuge: Community Reconstruction, Place Remaking, and Environmental Justice in the City*. Urban and Industrial Environments. MIT Press.

Anguelovski, Isabelle, James J. T. Connolly, Hamil Pearsall, Galia Shokry, Melissa Checker, Juliana Maantay, Kenneth Gould, Tammy Lewis, Andrew Maroko, and J. Timmons Roberts. 2019. "Opinion: Why Green 'Climate Gentrification' Threatens Poor and Vulnerable Populations." *Proceedings of the National Academy of Sciences* 116 (52): 26139–43. https://doi.org/10.1073/pnas.1920490117.

Aning, Agya K. 2021. "Baltimore Continues Incinerating Trash, Despite Opposition from Its New Mayor and City Council." *Inside Climate News* (blog), February 19. https://insideclimatenews.org/news/19022021/baltimore-continues-incinerating-trash-despite-opposition-from-its-new-mayor-and-city-council/.

Armstrong, Donna. 2000. "A Survey of Community Gardens in Upstate New York: Implications for Health Promotion and Community Development." *Health & Place* 6 (4): 319–27. https://doi.org/10.1016/S1353-8292(00)00013-7.

Asayama, Shinichiro, and Atsushi Ishii. 2017. "Selling Stories of Techno-Optimism? The Role of Narratives on Discursive Construction of Carbon Capture and Storage in the Japanese Media." *Energy Research & Social Science*, Narratives and Storytelling in Energy and Climate Change Research, 31 (September): 50–59. https://doi.org/10.1016/j.erss.2017.06.010.

Bailey, Anna, Sheelagh Barron, Costanza Curro, and Elizabeth Teague. 2018. "The System Made Me Do It: Strategies of Survival." In *Global Encyclopaedia of Informality* (Vol. 2) edited by Alena Ledeneva, 7–180. Understanding Social and Cultural Complexity. UCL Press. https://doi.org/10.2307/j.ctt20krxgs.9.

Banks, Charles, Sonia Heaven, Yue Zhang, and Urs Baier. 2018. *Food Waste Digestion: Anaerobic Digestion of Food Waste for a Circular Economy*. Edited by Jerry D Murphy. IEA Bioenergy Task 37.

Barad, Karen Michelle. 2007. *Meeting the Universe Halfway: Quantum Physics and the Entanglement of Matter and Meaning*. Duke University Press.

Barbanel, Josh. 1985. "As Thousands Protest, City Backs Trash Plant." *The New York Times*, May 23.

Barrett, Dawson. 2013. "DIY Democracy: The Direct Action Politics of U.S. Punk Collectives." *American Studies* 52 (2): 23–42. https://doi.org/10.1353/ams.2013.0039.

Bart, Pauline B. 1987. "Seizing the Means of Reproduction: An Illegal Feminist Abortion Collective—How and Why It Worked." *Qualitative Sociology* 10 (4): 339–57. https://doi.org/10.1007/BF00988383.

Belda-Miquel, Sergio, Victoria Pellicer-Sifres, and Alejandra Boni. 2020. "Exploring the Contribution of Grassroots Innovations to Justice: Using the Capability Approach to Normatively Address Bottom-Up Sustainable Transitions Practices." *Sustainability* 12 (9): 3617. https://doi.org/10.3390/su12093617.

Bennhold, Katrin. 2013. "Britain's Ministry of Nudges." *The New York Times*, December 7.

Bergmann, Frithjof. 2019. *New Work, New Culture: Work We Want and a Culture That Strengthens Us*. Zero Books.

Berlant, Lauren Gail. 2011. *Cruel Optimism*. Duke University Press.

Bierbaum, Tim, dir. 2015. "*The Settlers of Brooklyn*." Above Average. March 24, 2015. *Comedy sketch, 1:17*. https://www.youtube.com/watch?v=D-thbQlaMIc.

Bilsens Brolis, Linda. 2018. "Community-Scale Composting at Urban Gardens and Farms." *BioCycle*, January.

BioCycle. 2014. "Composting Facility Ordered to Close," November.

BioCycle CONNECT. 2020. "New York City Council Restores Funding for NYC Compost Project," July 8. https://www.biocycle.net/new-york-city-council-restores-funding-for-nyc-compost-project/.

BK ROT. 2016. "BK ROT : Bios." https://www.bkrot.org/bios.

BK ROT. 2020. "Compost Intention." https://www.bkrot.org/product-page/compost-intention. Archived at https://web.archive.org/web/20201202144924/https://www.bkrot.org/product-page/compost-intention.

BK ROT (@bk_rot). 2021. "A reminder that compositing is an ancestral practice that spans millennia." June 3, 2021. https://www.instagram.com/p/CPqrIy2DGkp/?hl=en

Block, Dave, and Nora Goldstein. 2000. "Solid Waste Composting Trends in the U.S." *BioCycle*, November.

Blum, Barton. 1992. "Composting and the Roots of Sustainable Agriculture." *Agricultural History* 66 (2): 171–88.

Botetzagias, Iosif, Maria Kevrekidou, Chrisovalandis Malesios, and Nikoleta Jones. 2020. "Exercising Social Control in PAYT (Pay-As-You-Throw)

Violations: The Role of Subjective Evaluations and Social Capital." *Waste Management* 105 (15): 347–54.

Bowker, Geoffrey C., and Susan Leigh Star. 2008. *Sorting Things Out: Classification and Its Consequences*. MIT Press.

Bradford, Abigail, Jonathan Sundby, Alexander Truelove, and Adair Andre. 2019. "Composting in America: A Path to Eliminate Waste, Revitalize Soil and Tackle Global Warming." US PIRG. https://uspirg.org/sites/pirg/files/reports/USP%20Composting%20in%20America%20FINAL.pdf.

Bradley, Karin. 2018. "Bike Kitchens—Spaces for Convivial Tools." *Journal of Cleaner Production*, Technology and Degrowth, 197 (October): 1676–83. https://doi.org/10.1016/j.jclepro.2016.09.208.

Brown, Sally. 2012. "Connections: Almost Magical Microbes." *BioCycle*, September. https://www.biocycle.net/connections-almost-magical-microbes/.

Brown, Wendy. 2017. *Undoing the Demos: Neoliberalism's Stealth Revolution*. Near Futures. Zone Books.

Bulkeley, Harriet, Lars Coenen, Niki Frantzeskaki, Christian Hartmann, Annica Kronsell, Lindsay Mai, Simon Marvin, Kes McCormick, Frank van Steenbergen, and Yuliya Voytenko Palgan. 2016. "Urban Living Labs: Governing Urban Sustainability Transitions." *Current Opinion in Environmental Sustainability* 22 (October): 13–17. https://doi.org/10.1016/j.cosust.2017.02.003.

Bullard, Robert D. 1990. *Dumping in Dixie: Race, Class, and Environmental Quality*. Westview Press.

Bullard, Robert, and Beverly Hendrix Wright. 1987. "Environmentalism and the Politics of Equity: Emergent Trends in the Black Community." *Mid-American Review of Sociology* 12 (2): 21–37.

Bureau of Labor Statistics. 2021. "National Census of Fatal Occupational Injuries in 2020." USDL-21-2145. Bureau of Labor Statistics.

Buzby, Jean C., Hodan Farah Wells, Bruce Axtman, and Jana Mickey. 2009. "Supermarket Loss Estimates for Fresh Fruit, Vegetables, Meat, Poultry, and Seafood and Their Use in the ERS Loss-Adjusted Food Availability Data." 44. Economic Information Bulletin. USDA. https://www.ers.usda.gov/publications/pub-details/?pubid=44309.

Cahill, Caitlin. 2007. "Negotiating Grit and Glamour: Young Women of Color and the Gentrification of the Lower East Side." *City & Society* 19 (2): 202–31. https://doi.org/10.1525/city.2007.19.2.202.

Campbell, Stu. 1998. *Let It Rot! The Gardener's Guide to Composting*. 3rd. ed. Storey's down-to-Earth Guides. Storey Communications.

Carpenter, Susan. 2014. "The 'Final Frontier' of Recycling." *Orange County Register*, July 16.

Carver, George Washington. 1917. *Twelve Ways to Meet the New Economic*

Conditions Here in the South. Tuskegee Normal and Industrial Institute.

———. 1936. *How to Build up and Maintain the Virgin Fertility of Our Soils*. Tuskegee Institute Press. http://archive.org/details/CAT31355516.

Casey, John, and Apurva Mehrotra. 2011. "Solid Waste Management and Environmental Justice: Building and Sustaining Coalitions." Baruch College.

Cassels, Louis. 1971. "Recycling May Convert Trash to Economic Asset." *Beaver County Times*, March 26.

Chakravarty, Surajit, and Felicity Hwee-Hwa Chan. 2016. "Imagining Shared Space: Multivalent Murals in New Ethnic '-Towns' of Los Angeles." *Space and Culture* 19 (4): 406–20. https://doi.org/10.1177/1206331215621018.

Champeny, Ana. 2018. "Update on the City's Organics Collection Program." Citizens Budget Commission of New York. September 20. https://cbcny.org/advocacy/update-citys-organics-collection-program.

Chan, Joana, Bryce DuBois, and Keith G. Tidball. 2015. "Refuges of Local Resilience: Community Gardens in Post-Sandy New York City." *Urban Forestry & Urban Greening* 14 (3): 625–35. https://doi.org/10.1016/j.ufug.2015.06.005.

Chancey, Bahij V. 2021. "Community Composting: Public-Nonprofit Partnerships and Equity in New York City Organic Waste Programs." Thesis, Massachusetts Institute of Technology.

Checker, Melissa. 2011. "Wiped Out by the 'Greenwave': Environmental Gentrification and the Paradoxical Politics of Urban Sustainability." *City & Society* 23 (2): 210–29. https://doi.org/10.1111/j.1548-744X.2011.01063.x.

Chien, Ker-hsuan. 2018. "Entrepreneurialising Urban Informality: Transforming Governance of Informal Settlements in Taipei." *Urban Studies* 55 (13): 2886–902. https://doi.org/10.1177/0042098017726739.

Citizens Budget Commission. 2016. "Can We Have Our Cake and Compost It Too?" https://cbcny.org/research/can-we-have-our-cake-and-compost-it-too.

City of Cambridge. 2019. "Curbside Composting." March 23. https://www.cambridgema.gov:443/en/Services/curbsidecomposting.

City of New York, Mayor's Office. 2013a. "Mayor Bloomberg Announces More Than 100 Restaurants to Participate in the 'Food Waste Challenge,' New City Program to Reduce Amount of Waste Sent to Landfills." The Official Website of the City of New York, April 25. http://www1.nyc.gov/office-of-the-mayor/news/148-13/mayor-bloomberg-more-100-restaurants-participate-the-food-waste-challenge-.

———. 2013b. "Mayor Bloomberg Launches New 'Recycle Everything' Ads and Announces Expansion of the Organic Food Waste Recycling Program." July 29, 2013. https://www1.nyc.gov/office-of-the-mayor/news

/260-13/mayor-bloomberg-launches-new-recycle-everything-ads-announces-expansion-the-organic-food.
———. 2014. "#BrownBin: The de Blasio Family on Food and Waste Recycling." NYC Mayor's Office. August 6, 2014. Public service announcement, 1:01. https://www.youtube.com/watch?v=GBeA53ZsumY.
———. 2021. "Transcript: Mayor de Blasio Holds Media Availability." Office of the Mayor. April 22, 2021. http://www1.nyc.gov/office-of-the-mayor/news/295-21/transcript-mayor-de-blasio-holds-media-availability.
City of New York, Mayor's Office of Long Term Planning and Sustainability. 2007. "PlaNYC: A Greener, Greater New York." New York. https://www.nyc.gov/html/planyc/downloads/pdf/publications/full_report_2007.pdf.
City of New York, Mayor's Office of Sustainability. 2015. "One New York: The Plan for a Strong and Just City." Office of Long-Term Planning and Sustainability. http://www.nyc.gov/html/onenyc/downloads/pdf/publications/OneNYC.pdf.
City of Toronto. 2017. "What Goes in the Green Bin?" https://www.toronto.ca/wp-content/uploads/2018/01/95aa-98d6-A1700930_WhatGoesInTheGreenBinGuide_AODA.pdf.
Clark, Nate. 2015. "The Business of Community Composting." *BioCycle*, January.
Clarke, Marjorie J. 1993. "Integrated Municipal Solid Waste Planning and Decision-Making in New York City: The Citizens' Alternative Plan." *Air & Waste* 43 (4): 453–62. https://doi.org/10.1080/1073161X.1993.10467143.
Coker, Craig. 2021. "Spontaneous Combustion in Composting: The Causes." *BioCycle CONNECT*, July 14. https://www.biocycle.net/spontaneous-combustion-in-composting-the-causes/.
Coleman-Jensen, Alisha, Christian Gregory, and Anita Singh. 2014. "Household Food Security in the United States in 2013." USDA. https://www.ers.usda.gov/webdocs/publications/45265/48787_err173.pdf.
Collins, H. M., and Robert Evans. 2002. "The Third Wave of Science Studies of Expertise and Experience." *Social Studies of Science* 32 (2): 235–96. https://doi.org/10.1177/0306312702032002003.
Collins, Lisa M. 2018. "The Pros and Cons of New York's Fledgling Compost Program." *The New York Times*, November 9.
Colombijn, Freek. 2020. "Secrecy at the End of the Recycling Chain: The Recycling of Plastic Waste in Surabaya, Indonesia." *Worldwide Waste: Journal of Interdisciplinary Studies* 3 (1): 2. https://doi.org/10.5334/wwwj.43.
Committee on Sanitation and Solid Waste Management (New York City Council). 2013. Committee Report November 22. "A Local Law to Amend the Administrative Code of the City of New York, in Relation

to Commercial Organic Waste. (Int. No. 1162)." https://legistar.council.nyc.gov/View.ashx?M=F&ID=2735045&GUID=745B4F1F-4B81-4DBA-88A3-21EEFA85EAA8.

———. 2018. *Executive Budget Hearings—Sanitation and Solid Waste Management.* May 17. New York City Legistar. https://legistar.council.nyc.gov/LegislationDetail.aspx?ID=3490250&GUID=5D437490-C1D4-42AF-9DF0-F845A74EB167&Options=&Search=.

———. 2020. *Organic Waste Drop off Sites.* June 15. New York City Legistar. https://legistar.council.nyc.gov/View.ashx?M=F&ID=8640443&GUID=FC630F6D-C0C6-4624-B6D6-02E2ACFD1F07.

Committee on Sanitation and Solid Waste Management (New York City Council) jointly with the Committee on Finance. 2021. *Executive Budget Hearings—Sanitation and Solid Waste Management.* May 14. New York City Legistar. https://legistar.council.nyc.gov/View.ashx?M=F&ID=9550552&GUID=D779E840-D0BD-4AA2-B1D3-90727C7376C2.

Connell, Raewyn, Barbara Fawcett, and Gabrielle Meagher. 2009. "Neoliberalism, New Public Management and the Human Service Professions: Introduction to the Special Issue." *Journal of Sociology* 45 (4): 331–38. https://doi.org/10.1177/1440783309346472.

Corburn, Jason. 2005. *Street Science: Community Knowledge and Environmental Health Justice.* MIT Press.

Cornell Waste Management Institute. 1996. "The Science and Engineering of Composting." http://compost.css.cornell.edu/science.html.

Covert, Matt, and Alfonso Morales. 2014. "Formalizing City Farms: Conflict and Conciliation." In *The Informal American City: Beyond Taco Trucks and Day Labor*, edited by Vinit Mukhija and Anastasia Loukaitou-Sideris, 193–208. MIT Press.

Crunden, E. A. 2020. "New York Follows through with Sanitation Cuts, Preserves Some Organics Funding | Waste Dive." *Waste Dive*, May 21. https://www.wastedive.com/news/new-york-organics-suspension-compost-coronavirus-pushback-lawmakers/578036/.

Cruz, Jesi Taylor. 2020. "How I Have Hope: Utopian Thought." *Hey Alma* (blog), November 3. https://www.heyalma.com/how-i-have-hope-utopian-thought/.

Curran, Winifred, and Trina Hamilton. 2012. "Just Green Enough: Contesting Environmental Gentrification in Greenpoint, Brooklyn." *Local Environment* 17 (9): 1027–42. https://doi.org/10.1080/13549839.2012.729569.

Cvetkovich, Ann. 2012. *Depression: A Public Feeling.* Duke University Press Books.

Delany, Samuel R. 2012. *Starboard Wine: More Notes on the Language of Science Fiction.* Rev. ed. Wesleyan University Press.

Deseret News. 1990. "Makers of Disposable Diapers Cry Foul." Deseret News. June 14. https://www.deseret.com/1990/6/14/18866358/makers-of-disposable-diapers-cry-foul.

Desfor, Gene, and Roger Keil. 2004. *Nature and the City: Making Environmental Policy in Toronto and Los Angeles*. University of Arizona Press.

Dillon, Lindsey. 2014. "Race, Waste, and Space: Brownfield Redevelopment and Environmental Justice at the Hunters Point Shipyard: Waste, Race and Space." *Antipode* 46 (5): 1205–21. https://doi.org/10.1111/anti.12009.

Dillon, Lindsey, and Julie Sze. 2018. "Equality in the Air We Breathe: Police Violence, Pollution, and the Politics of Sustainability." In *Sustainability: Approaches to Environmental Justice and Social Power*, edited by Julie Sze, 246–70. NYU Press.

DiSalvo, Carl. 2009. "Design and the Construction of Publics." *Design Issues* 25 (1): 48–63.

DNLee. 2011. "Urban Environmentalism in Action: St. Louis North City Residents Sue City over Recycling." *Urban Science Adventures* (blog), April 5. http://urban-science.blogspot.com/2011/04/urban-environmentalism-in-action-st.html.

———. 2016. "I Revisit My Thoughts on African-American Engagement in Environmentalism." Scientific American Blog Network. October 11. https://blogs.scientificamerican.com/urban-scientist/i-revisit-my-thoughts-on-african-american-engagement-in-environmentalism/.

Doersam, Jim, and Paul Gregory. 2016. "Public/Private Partnership Creates Composting Infrastructure." *BioCycle*, April.

DSNY. 2006. "Comprehensive Solid Waste Management Plan." New York City Department of Sanitation. www1.nyc.gov (no longer available).

———. 2012. "Local Law 42: A 2012 Assessment of Composting Opportunities." NYC Department of Sanitation.

———. 2014. "2014 NYC Community Composting Report." DSNY. http://www1.nyc.gov/assets/dsny/docs/about_2014-community-composting-report-LL77_0815.pdf.

———. 2015a. "2015 NYC Organics Collection Report." DSNY. http://www1.nyc.gov/assets/dsny/downloads/pdf/studies-and-reports/OrganicsCollection-LL77-NYCOrganicsCollectionReport-2015.pdf.

———. 2015b. *NYC Master Composter Manual*. New York. https://www.nyc.gov/assets/dsny/docs/nyc-master-composter-manual-mcm.pdf.

———. 2016. "NYC Achievement: NYC Will Have Largest Organic Waste Curbside Collection Program in Nation." https://www1.nyc.gov/assets/dsny/site/resources/press-releases/nyc-achievement-nyc-will-have-largest-organic-waste-curbside-collection-program-in-nation.

———. 2017. "NYC Residential, School, and NYCHA Waste Characterization Study." New York, NY. https://dsny.cityofnewyork.us/wp-content/uploads/2018/04/2017-Waste-Characterization-Study.pdf.

———. 2021. "Annual Report New York City Municipal Refuse and Recycling Statistics: Fiscal Year 2020." https://dsny.cityofnewyork.us/wp-content/uploads/2020/09/about_dsny-non-dsny-collections-FY2020.pdf.

———. 2022. "Ways to Compost in NYC." Make Compost, Not Trash. August 4. https://www.makecompost.nyc/ (no longer available).

———. 2023. "DSNY Monthly Tonnage Data." NYC OpenData. https://data.cityofnewyork.us/City-Government/DSNY-Monthly-Tonnage-Data/ebb7-mvp5.

———. n.d. "Ways to Compost in NYC." Nature Materials. https://www.makecompost.nyc/compost. Archived at https://web.archive.org/web/20220822162652/https://www.makecompost.nyc/compost.

Duggan, Kevin. 2021. "New Yorkers Not Ready for Mandatory Compost Recycling: De Blasio | AmNewYork." *AMNY*, April 23. https://www.amny.com/news/new-york-compost-not-ready-de-blasio/.

"Dystopian Events Repackaged as Feel-Good Stories Are Really Getting Old [Facebook Group]." 2019. Facebook. https://www.facebook.com/groups/428242431084635/.

Earth Matter. 2014. "Short History of Compost." https://earthmatter.org/wp-content/uploads/2014/04/History-of-Compost.pdf.

Edwards, Ferne, and Dave Mercer. 2012. "Food Waste in Australia: The Freegan Response." *The Sociological Review* 60 (December): 174–91. https://doi.org/10.1111/1467-954X.12044.

Edwards, Jessy. 2022. "BK Pols Rally against Sanitation Budget Cuts amid Spike in Rats, Trash." *BK Reader* (blog), March 4. https://bkreader.com/2022/03/04/bk-pols-rally-against-sanitation-budget-cuts-as-streets-get-dirtier-rattier/.

Elias, Norbert. 2000. *The Civilizing Process: Sociogenetic and Psychogenetic Investigations*. Rev. ed. Blackwell Publishing.

Eliasoph, Nina. 2013. *The Politics of Volunteering*. Political Sociology Series. Polity Press.

Elsheshtawy, Yasser. 2011. "Informal Encounters: Mapping Abu Dhabi's Urban Public Spaces." *Built Environment (1978–)* 37 (1): 92–113.

Epstein, Eliot. 1996. *The Science of Composting*. CRC Press.

Epstein, Steven. 1996. *Impure Science: AIDS, Activism, and the Politics of Knowledge*. University of California Press.

Fabricant, Nicole. 2022. "Starve the Beast: Community-Owned and Community-Controlled Composting as an Alternative to Incineration in Baltimore." *Metropolitics*, June. https://metropolitics.org/Starve-the-Beast-Community-Owned-and-Community-Controlled-Composting-as-an.html.

Feldman, Kiera. 2018. "Trashed: Inside the Deadly World of Private Garbage Collection." *ProPublica*. https://www.propublica.org/article/trashed-inside-the-deadly-world-of-private-garbage-collection.

Ferrell, John S. 2010. "George Washington Carver: A Blazer of Trails to a Sustainable Future." In *Land and Power: Sustainable Agriculture and African Americans: A Collection of Essays from the 2007 Black Environmental Thought Conference*, edited by Jeffrey L. Jordan, Edward Pennick, Walter Hill, and Robert Zabawa, 11–32. Sustainable Agriculture Research and Education Program.

Ferro, John. 2016. "Food Waste: Recycling's Next Frontier." *Poughkeepsie Journal*, August 6. https://www.poughkeepsiejournal.com/story/tech/science/environment/2016/08/06/recycling-food-waste-dutchess-ulster/87572290/.

Fiore, Anthony. 2015. "Food-Energy-Water Nexus." Presented at NYC Organics Is Expanding into CB1, Brooklyn, NY, June 3.

Firth, Chris, Damian Maye, and David Pearson. 2011. "Developing 'Community' in Community Gardens." *Local Environment* 16 (6): 555–68. https://doi.org/10.1080/13549839.2011.586025.

Fishbein, Rebecca. 2015. "Greenpoint Residents Say Composting Is Stinking Up the Neighborhood." *Gothamist (blog)*, July 18.

Fisher, Mark. 2009. *Capitalist Realism: Is There No Alternative?* Zero Books.

Fitzmahan, Kaitlin. 2013. "Exploring Community Murals as a Tool for Community Development: A Case Study of Precita Eyes Muralists and the 24th Street/Mission District." Dissertation, University of California–Davis. http://gradworks.umi.com/15/46/1546197.html.

Flachs, Andrew. 2010. "Food for Thought: The Social Impact of Community Gardens in the Greater Cleveland Area." *Electronic Green Journal* 1 (30). https://doi.org/10.5070/G313010826.

Foote, Stephanie, and Elizabeth Mazzolini, eds. 2012. *Histories of the Dustheap: Waste, Material Cultures, Social Justice*. Urban and Industrial Environments. MIT Press.

Forgrieve, Janet. 2014. "Can Companies Cut Food Waste and Feed the World?" *SmartBlog on Food & Beverage* (blog), January 6. https://corp.smartbrief.com/original/2016/05/can-companies-cut-food-waste-and-feed-world-2.

Forlano, Laura. 2016. "Hacking the Feminist Disabled Body: The Journal of Peer Production." *Journal of Peer Production*, no. 8. http://peerproduction.net/issues/issue-8-feminism-and-unhacking/peer-reviewed-papers/hacking-the-feminist-disabled-body/.

Fortun, Kim. 2001. *Advocacy after Bhopal: Environmentalism, Disaster, New Global Orders*. 1 edition. University Of Chicago Press.

Foster, Ellen K. 2015. "The Politics of No Politics: Exploring Gender, Skill-Sharing, and Community-Building in Maker Communities." Presented at the Society for the Social Studies of Science, Denver, CO, November 14.

———. 2019. "Claims of Equity and Expertise: Feminist Interventions in the Design of DIY Communities and Cultures." *Design Issues* 35 (4): 33–41. https://doi.org/10.1162/desi_a_00562.

Foster, Ellen K., and Yana Boeva. 2018. "Making Sense of Place: Research Reflections from Two Multi-Sited Ethnographies." *Making Futures* 5. Conference paper.

Foster, John Bellamy. 1999. "Marx's Theory of Metabolic Rift: Classical Foundations for Environmental Sociology." *American Journal of Sociology* 105 (2): 366–405. https://doi.org/10.1086/210315.

Foster, John Bellamy, and Fred Magdoff. 1998. "Liebig, Marx and the Depletion of Soil Fertility: Relevance for Today's Agriculture." *Monthly Review* 50 (3): 32. https://doi.org/10.14452/MR-050-03-1998-07_3.

Foucault, Michel. 2010. *The Birth of Biopolitics: Lectures at the Collège de France, 1978—79*. Edited by Michel Senellart. 1st pbk. Ed. [Repr.]. Lectures at the Collège de France. Picador.

Fox, Sarah, Rachel Rose Ulgado, and Daniela Rosner. 2015. "Hacking Culture, Not Devices: Access and Recognition in Feminist Hackerspaces." In *Proceedings of the 18th ACM Conference on Computer Supported Cooperative Work & Social Computing*, 56–68. CSCW '15. ACM. https://doi.org/10.1145/2675133.2675223.

Fredericks, Rosalind. 2014. "Vital Infrastructures of Trash in Dakar." *Comparative Studies of South Asia, Africa and the Middle East* 34 (3): 532–48. https://doi.org/10.1215/1089201X-2826085.

Freeman, Juri, Lisa Skumatz, and Dana D'Souza. 2012. "Bolder County Compost Market Study." Economic Research Associates. https://assets.bouldercounty.org/wp-content/uploads/2017/02/RCD_CompostMarketStudy.pdf.

Furfaro, Danielle. 2015. "Wake Up and Smell the Rotting Coffee Grounds!" *Brooklyn Paper*, July 15. http://www.brooklynpaper.com/stories/38/29/dtg-greenpoint-composting-stinks-2015-07-17-bk.html.

Galbraith, Kate. 2009. "Composting Dirty Diapers in Toronto." *Green* (blog), *New York Times*. April 27, 2009.

Galison, Peter. 1987. *How Experiments End*. University of Chicago Press.

Gardezi, Maaz, and J. Gordon Arbuckle. 2020. "Techno-Optimism and Farmers' Attitudes Toward Climate Change Adaptation." *Environment and Behavior* 52 (1): 82–105. https://doi.org/10.1177/0013916518793482.

Geels, Dr. Ir. F. W. 2005. "The Dynamics of Transitions in Socio-Technical Systems: A Multi-Level Analysis of the Transition Pathway from Horse-Drawn Carriages to Automobiles (1860–1930)." *Technology Analysis & Strategic Management* 17 (4): 445–76. https://doiorg/10.1080/09537320500357319.

Geels, Frank W. 2011. "The Multi-Level Perspective on Sustainability

Transitions: Responses to Seven Criticisms." *Environmental Innovation and Societal Transitions* 1 (1): 24–40. https://doi.org/10.1016/j.eist.2011.02.002.

Ghose, Rina, and Margaret Pettygrove. 2014. "Urban Community Gardens as Spaces of Citizenship." *Antipode* 46 (4): 1092–112. https://doi.org/10.1111/anti.12077.

Gibson-Graham, J. K. 2006. *A Postcapitalist Politics*. University of Minnesota Press.

Gibson-Graham, J. K., J. Cameron, K. Dombroski, S. Healy, E. Miller, and the Community Economies Collective. 2020. "Cultivating Community Economies." Next System Project. https://www.communityeconomies.org/publications/popular-writing/cultivating-community-economies.

Gieryn, Thomas F. 1999. *Cultural Boundaries of Science: Credibility on the Line*. University of Chicago Press.

Gille, Zsuzsa. 2007. *From the Cult of Waste to the Trash Heap of History: The Politics of Waste in Socialist and Postsocialist Hungary*. Indiana University Press.

Gilson, Dave, and Kiera Butler. 2008. "A Brief History of the Disposable Diaper." *Mother Jones*, June 2008. https://www.motherjones.com/environment/2008/04/brief-history-disposable-diaper/.

Goldenberg, Sally, and Danielle Muoio Dunn. 2020. "Wasted Potential: The Consequences of New York City's Recycling Failure." *Politico*, January 5. https://www.politico.com/states/new-york/albany/story/2020/01/05/wasted-potential-the-consequences-of-new-york-citys-recycling-failure-1243578.

Goldman, Joanne Abel. 1997. *Building New York's Sewers: Developing Mechanisms of Urban Management*. Purdue University Press.

Goldstein, Eric. 2014. "New York City's History-Making Recycling Law Turns 25 Years Old (Part I)." *Natural Resources Defense Council* (blog), July 29. https://www.nrdc.org/bio/eric-goldstein/new-york-citys-history-making-recycling-law-turns-25-years-old-part-i.

———. 2015. "De Blasio Administration to Require Composting by NYC's Largest Food Waste Generators." *Natural Resources Defense Council (blog)*, July 2. https://www.nrdc.org/bio/eric-goldstein/blasio-administration-require-composting-nycs-largest-food-waste-generators.

Goldstein, Jesse. 2019. "From Planetary Improvement to Energy Abolition." Presented at the Engineering the Anthropocene, Troy, NY, December 11.

Goldstein, Joshua. 2021. *Remains of the Everyday: A Century of Recycling in Beijing*. University of California Press.

Goldstein, Nora. 1999. "Saving the Soil." *BioCycle*, November 1999.

———. 2005. "Mixed MSW Composting Facilities in the U.S." *BioCycle*, November.

———. 2009a. "Compost Science: Journal of Advocacy and Foresight." *BioCycle*, April.

———. 2009b. "Evolution of the Compost Windrow Turner." *BioCycle*, July.

———. 2017a. "The State of Organics Recycling in the U.S." The J. G. Press, Inc.

———. 2017b. "Big Apple Goes Big on Organics Recycling." *BioCycle*, January.

Goldstein, Nora, and Robert Steuteville. 1994. "Solid Waste Composting Seeks Its Niche." *BioCycle*, November.

Golueke, Clarence. 2009. "Composting Persepctives—Progress since 1950." *BioCycle*, May.

Golueke, C. G., and L. F. Diaz. 1996. "Historical Review of Composting and Its Role in Municipal Waste Management." In *The Science of Composting*, edited by Marco de Bertoldi, Paolo Sequi, Bert Lemmes, and Tiziano Papi, 3–14. Springer Netherlands. https://doi.org/10.1007/978-94-009-1569-5_1.

Goodyear, Sarah. 2015. "The Backlash to Gentrification Has Inspired Its Own Backlash." *CityLab* (blog), October 1. http://www.citylab.com/politics/2015/10/gentrification-backlash-has-inspired-its-own-backlash/408388/.

Gordon, Avery. 2004. *Keeping Good Time: Reflections on Knowledge, Power, and People*. Paradigm Publishers.

Gorrie, Peter. 2015. "Compost Use Trials on Ontario Farms." *BioCycle*, June.

Gould, Kenneth Alan, and Tammy L. Lewis. 2016. *Green Gentrification: Urban Sustainability and the Struggle for Environmental Justice*. Routledge Equity, Justice and the Sustainable City Series. Routledge.

Grandoit, Alice. 2020. "Adrienne Maree Brown on Creating the Future." *Deem*. https://www.deemjournal.com/stories/amb.

Greenberg, Miriam. 2008. *Branding New York: How a City in Crisis Was Sold to the World*. Taylor & Francis Group.

———. 2013. "What on Earth Is Sustainable? Toward Critical Sustainability Studies." *Boom* 3 (4): 54–66. https://doi.org/10.1525/boom.2013.3.4.54.

———. 2015. "'The Sustainability Edge': Competition, Crisis, and the Rise of Green Urban Branding." In *Sustainability in the Global City: Myth and Practice*, edited by Cindy Isenhour, Gary McDonogh, and Melissa Checker, 105–30. New Directions in Sustainability and Society. Cambridge University Press. https://doi.org/10.1017/CBO9781139923316.008.

Greenpointers. 2020. "Food Scrap Drop-Off Returns to McCarren Greenmarket." *Greenpointers* (blog), November 13. https://greenpointers.com/2020/11/13/food-scrap-drop-off-returns-to-mccarren-greenmarket/.

Greer, Diane. 2012. "Commercial Food Waste Recovery in New York City." *BioCycle*, December.

Gross, Deborah, James Levy, David Pinney, and Kathleen Schomaker. 1997. "New Milford Farms and Organic Residue Recycling." *Yale E&ES Bulletin*, no. 106:49–67.

Gross, Joan. 2009. "Capitalism and Its Discontents: Back-to-the-Lander and

Freegan Foodways in Rural Oregon." *Food and Foodways* 17 (2): 57–79. https://doi.org/10.1080/07409710902925797.

"Guano as a Manure." 1852. *Scientific American* 8 (10): 80–80. https://doi.org/10.1038/scientificamerican11201852-80.

Guevarra, Leslie. 2011. "Waste Management Takes Two Big Steps in the Com-posting Business." *GreenBiz* (blog), May 6. https://www.greenbiz.com/news/2011/05/06/waste-management-takes-two-big-steps-com posting-business.

Gunders, Dana. 2012. "Wasted: How America Is Losing Up to 40 Percent of Its Food from Farm to Fork to Landfill." National Resources Defense Council. http://www.nrdc.org/food/files/wasted-food-ip.pdf.

Guston, David H. 2001. "Boundary Organizations in Environmental Policy and Science: An Introduction." *Science, Technology, & Human Values* 26 (4): 399–408. https://doi.org/10.1177/016224390102600401.

Guthman, Julie. 2004. *Agrarian Dreams: The Paradox of Organic Farming in California*. University of California Press.

———. 2008a. *The Unbearable Whiteness of Alternative Food*. MP3. Food Politics. Madison. http://www.havenscenter.org/ (no longer available).

———. 2008b. "'If They Only Knew': Color Blindness and Universalism in California Alternative Food Institutions." *The Professional Geographer* 60 (3): 387–97. https://doi.org/10.1080/00330120802013679.

———. 2008c. "Bringing Good Food to Others: Investigating the Subjects of Alternative Food Practice." *Cultural Geographies* 15 (4): 431–47. https://doi.org/10.1177/1474474008094315.

Guttmann, E. B. A. 2005. "Midden Cultivation in Prehistoric Britain: Arable Crops in Gardens." *World Archaeology* 37 (2): 224–39. https://doi.org/10.1080/00438240500094937.

Hackworth, Jason, and Neil Smith. 2001. "The Changing State of Gentrification." *Tijdschrift Voor Economische En Sociale Geografie* 92 (4): 464–77. https://doi.org/10.1111/1467-9663.00172.

Hamilton, Jennifer Mae, and Astrida Neimanis. 2018. "Composting Feminisms and Environmental Humanities." *Environmental Humanities* 10 (2): 501–27. https://doi.org/10.1215/22011919-7156859.

Haraway, Donna Jeanne. 2008. *When Species Meet*. Posthumanities 3. University of Minnesota Press.

———. 2016. *Staying with the Trouble: Making Kin in the Chthulucene*. Experimental Futures: Technological Lives, Scientific Arts, Anthropological Voices. Duke University Press.

Hargreaves, Tom, Alex Haxeltine, Noel Longhurst, and Gill Seyfang. 2011. "Sustainability Transitions from the Bottom-up: Civil Society, the Multi-Level Perspective and Practice Theory." *Working Paper—Centre for Social and Economic Research on the Global Environment* 1:1–26.

Harvey, David. 2007. "Neoliberalism as Creative Destruction." *The Annals of the American Academy of Political and Social Science* 610:22–44.

Hatch, Kendall. 2013. "Marlborough: Officials Sue Contractor." *WickedLocal.Com*, November 20. http://www.wickedlocal.com/article/20131120/News/311209372.

Hawkins, Gay. 2006. *The Ethics of Waste: How We Relate to Rubbish*. Rowman & Littlefield Publishers.

HDR (Hennington, Durham & Richardson). 2004. "Commercial Waste Management Study Volume II: Commercial Waste Generation and Projections." Omaha, NE. http://www1.nyc.gov/assets/dsny/downloads/pdf/studies-and-reports/2004-commercial-waste-management-study/v2-cwgp.pdf.

Hepperly, Paul, and Christine Ziegler Usch. 2007. "Ancient Wisdom Meets Modern Science: Studies & Advances in Composting Technology." *Acres U.S.A.*, September 2007.

Herrera, Lenny. 2015. "6 Serious Signs of Gentrification in New York City." *Odyssey*, November 8. http://theodysseyonline.com/suny-binghamton/5-signs-your-neighborhood-is-gentrifying/212342.

Hersey, Mark. 2006. "Hints and Suggestions to Farmers: George Washington Carver and Rural Conservation in the South." *Environmental History* 11 (2): 239–68.

Hersey, Mark D. 2011. *My Work Is That of Conservation: An Environmental Biography of George Washington Carver*. University of Georgia Press.

Hess, David J. 2004. "Medical Modernisation, Scientific Research Fields and the Epistemic Politics of Health Social Movements." *Sociology of Health & Illness* 26 (6):695–709. https://doi.org/10.1111/j.0141-9889.2004.00414.x.

———. 2007. *Alternative Pathways in Science and Industry: Activism, Innovation, and the Environment in an Era of Globalization*. MIT Press.

———. 2012. *Good Green Jobs in a Global Economy: Making and Keeping New Industries in the United States*. Urban and Industrial Environments. MIT Press.

Heynen, Nik. 2020. "'A Plantation Can Be a Commons': Re-Earthing Sapelo Island through Abolition Ecology: The 2018 Neil Smith Lecture." *Antipode* 53 (1): 95–114. https://doi.org/10.1111/anti.12631.

Heynen, Nik, and Megan Ybarra. 2021. "On Abolition Ecologies and Making 'Freedom as a Place.'" *Antipode* 53 (1): 21–35. https://doi.org/10.1111/anti.12666.

Hickman, H. Lanier. 2003. *American Alchemy: The History of Solid Waste Management in the United States*. Forester Press.

Hird, Myra J., Scott Lougheed, R. Kerry Rowe, and Cassandra Kuyvenhoven. 2014. "Making Waste Management Public (or Falling Back to Sleep)." *Social Studies of Science* 44 (3): 441–65. https://doi.org/10.1177/0306312713518835.

Hobbs, Allegra. 2015. "Noble Rot: Bushwick Community Garden Combines Compost Piles with Wild Flowers." *The Brooklyn Paper*, October 27,

2015. http://www.brooklynpaper.com/stories/38/44/dtg-know-waste-lands-wildlife-garden-2015-10-30-bk.html.

Holusha, John. 1991. "The Packaging Industry's New Fancy—Composting Garbage." *The New York Times*, February 24.

Howard, Albert. 2010. *An Agricultural Testament*. Oxford University Press.

Integrated Waste Management Board. 1992. "Food for Thought: Restaurant Guide to Waste Reduction and Recycling." Integrated Waste Management Board. https://p2infohouse.org/ref/01/00030.pdf.

Jacobs, Jane. 1961. *The Death and Life of Great American Cities*. Vintage Books.

Jacobson, Abbi and Ilana Glazer. *Broad City*. Season 1, episode 2, "Pu$$y Weed." Aired January 29, 2014, on Comedy Central.

Jalbert, Kirk, and Abby J. Kinchy. 2015. "Sense and Influence: Environmental Monitoring Tools and the Power of Citizen Science." *Journal of Environmental Policy & Planning* 18 (3): 379–97. https://doi.org/10.1080/1523908X.2015.1100985.

Jantz, Eric. 2017. "Environmental Racism with a Faint Green Glow." *Natural Resources Journal* 58 (2): 247.

Järvensivu, Paavo, Tero Toivanen, Tere Vadén, Ville Lähde, Antti Majava, and Jussi T. Eronen. 2019. "Global Sustainable Development Report 2019: Invited Background Document on Economic Transformation, to Chapter 'Transformation: The Economy.'" United Nations. https://bios.fi/bios-governance_of_economic_transition.pdf.

Jasanoff, Sheila, and Sang-Hyun Kim. 2009. "Containing the Atom: Sociotechnical Imaginaries and Nuclear Power in the United States and South Korea." *Minerva* 47 (2): 119–46.

Jensen, Casper Bruun, and Atsuro Morita. 2015. "Infrastructures as Ontological Experiments." *Engaging Science, Technology, and Society* 1 (November): 81–87. https://doi.org/10.17351/ests2015.21.

Johnston, Marsha W. 2018. "Extreme Composting at Standing Rock." *BioCycle*, September.

Jorgensen, Jillian. 2015. "'Save As You Throw' Working Group Will Consider Financial Incentive for Recycling." *Observer* (blog), *AP News*. November 19. https://observer.com/2015/11/save-as-you-throw-working-group-will-consider-financial-incentive-for-recycling/.

Karidis, Arlene. 2016. "NYC Will Triple Food Waste Collections by Extending Service to 600K Residences in 2017." *Waste Dive*, March 18. http://www.wastedive.com/news/nyc-will-triple-food-waste-collections-by-extending-service-to-600k-residen/415875/.

Keil, Jennifer Gould. 2014. "Rents Soar High in Bushwick as Popularity Grows." *New York Post*, September 12. http://nypost.com/2014/09/12/rents-soar-high-in-bushwick-as-popularity-grows/.

Keil, Roger. 2009. "The Urban Politics of Roll-with-it Neoliberalization." *City* 13 (2–3): 230–45. https://doi.org/10.1080/13604810902986848.

Keller, Evelyn Fox. 1983. *A Feeling for the Organism: The Life and Work of Barbara McClintock*. W. H. Freeman.

Kendall, Gavin, and Gary Wickham. 1999. *Using Foucault's Methods*. SAGE Publications Ltd. https://doi.org/10.4135/9780857020239.

Kendall, Tina. 2007. "Utopia Gleaners." *Alphabet City Trash*, 222–9. http://alphabet-city.org/issues/trash/articles/utopia-gleaners.

Kenner, Alison. 2020. "Scrapping the Workshop of the World: Civic Infrastructuring and the Politics of Late Industrial Governance." *Engaging Science, Technology, and Society* 6 (November): 514–33. https://doi.org/10.17351/ests2020.391.

Kenney, Martha. 2019. "Fables of Response-Ability: Feminist Science Studies as Didactic Literature." *Catalyst: Feminism, Theory, Technoscience* 5 (1): 1–39. https://doi.org/10.28968/cftt.v5i1.29582.

Kiersz, Andy, 2014. "Here's a Block-by-Block Look at Who's Making How Much across NYC's 5 Boroughs." *Business Insider*. December 11. http://www.businessinsider.com/new-york-city-income-maps-2014-12.

Kinchy, Abby, Kirk Jalbert, and Jessica Lyons. 2014. "What Is Volunteer Water Monitoring Good For? Fracking and the Plural Logics of Participatory Science." In *Fields of Knowledge: Science, Politics and Publics in the Neoliberal Age*, edited by Scott Frickel and David J. Hess, 259–89. Political Power and Social Theory. Emerald.

King, Donna Lee. 1995. *Doing Their Share to Save the Planet: Children and Environmental Crisis*. Rutgers University Press.

Köhler, Jonathan, Frank W. Geels, Florian Kern, Jochen Markard, Elsie Onsongo, Anna Wieczorek, Floortje Alkemade, et al. 2019. "An Agenda for Sustainability Transitions Research: State of the Art and Future Directions." *Environmental Innovation and Societal Transitions* 31 (June): 1–32. https://doi.org/10.1016/j.eist.2019.01.004.

Lange, Robert. 1999. "Backyard Composting in New York City: A Comprehensive Program Evaluation." DSNY. http://www1.nyc.gov/assets/dsny/docs/about_1999-backyard-composting_0815.pdf.

———. 2001. "Composting in New York City: A Complete Program History." New York City Department of Sanitation, Bureau of Waste Prevention, Reuse, and Recycling. http://www.nyc.gov/ (no longer available).

Larkin, Brian. 2013. "The Politics and Poetics of Infrastructure." *Annual Review of Anthropology* 42 (1): 327–43. https://doi.org/10.1146/annurev-anthro-092412-155522.

Latour, Bruno. 1993. *The Pasteurization of France*. Harvard University Press.

Laundromat Project. 2023. "The Laundromat Project." Laundromat Project. 2023. https://laundromatproject.org/.

LaWare, Margaret R. 1998. "Encountering Visions of Aztlan: Arguments for Ethnic Pride, Community Activism and Cultural Revitalization in Chicano Murals." *Argumentation and Advocacy* 34 (3): 140–53.

Lawless, Jill. 2015. "An East London 'Cereal Bar' Was Vandalized by

Protestors for Being a Sign of Gentrification." *Business Insider*, October 1. http://www.businessinsider.com/ap-in-divided-london-trendy-cereal-cafe-targeted-by-protesters-2015-10.

Levitas, Ruth. 1990a. "Educated Hope: Ernst Bloch on Abstract and Concrete Utopia." *Utopian Studies* 1 (2): 13–26.

———. 1990b. *The Concept of Utopia*. 1st ed. Utopianism and Communitarianism. Syracuse University Press.

Liboiron, Max. 2021. *Pollution Is Colonialism*. Duke University Press.

Ligon Paul, P. N. Mishra, and Votta Tom. 2000. "Waste Service Providers Become Resource Managers." *BioCycle*, April.

Linnitt, Carol. 2020. "'It's about Terrorizing Us': Environmental Racism Expert Draws Parallels with Police Brutality." *YES! Magazine* (blog), June 18. https://www.yesmagazine.org/environment/2020/06/18/police-brutality-environmental-racism.

Loorbach, Derk, Niki Frantzeskaki, and Flor Avelino. 2017. "Sustainability Transitions Research: Transforming Science and Practice for Societal Change." *Annual Review of Environment and Resources* 42 (1): 599–626. https://doi.org/10.1146/annurev-environ-102014-021340.

Lorde, Audre. 2007. *Sister Outsider: Essays and Speeches*. Crossing Press.

Loughran, Kevin. 2014. "Parks for Profit: The High Line, Growth Machines, and the Uneven Development of Urban Public Spaces." *City & Community* 13 (1): 49–68. https://doi.org/10.1111/cico.12050.

Lounsbury, Michael, Marc Ventresca, and Paul M. Hirsch. 2003. "Social Movements, Field Frames and Industry Emergence: A Cultural–Political Perspective on US Recycling." *Socio-Economic Review* 1 (1): 71–104. https://doi.org/10.1093/soceco/1.1.71.

Lucier, Cristina A., and Brian J. Gareau. 2015. "From Waste to Resources? Interrogating 'Race to the Bottom' in the Global Environmental Governance of the Hazardous Waste Trade." *Journal of World-Systems Research* 21 (2): 495–520. https://doi.org/10.5195/jwsr.2015.11.

MacBride, Samantha. 2011. *Recycling Reconsidered: The Present Failure and Future Promise of Environmental Action in the United States*. MIT Press.

———. 2015. "Source Separated Organics in Large New York City Apartment Buildings." *BioCycle*, January.

MacGregor, Finlay, Vasna Ramasar, and Kimberly A. Nicholas. 2017. "Problems with Firm-Led Voluntary Sustainability Schemes: The Case of Direct Trade Coffee." *Sustainability* 9 (4): 651. https://doi.org/10.3390/su9040651.

Madanipour, Ali. 2015. "Social Exclusion and Space." In *The City Reader*, edited by Richard T. LeGates and Frederic Stout, 158–65. Routledge.

Madeson, Frances. 2022. "Organize, Don't Agonize, Says Cooperation Jackson's Kali Akuno." *The Real News Network*. March 25. http://therealnews.com/organize-dont-agonize-says-cooperation-jacksons-kali-akuno.

Maniates, Michael F. 2001. "Individualization: Plant a Tree, Buy a Bike, Save the World?" *Global Environmental Politics* 1 (3): 31–52. https://doi.org/10.1162/152638001316881395.

Marchaim, Uri. 1992. "Biogas Processes for Sustainable Development." Rome: Food and Agriculture Organization of the United Nations.

Marres, Noortje. 2009. "Testing Powers of Engagement: Green Living Experiments, the Ontological Turn and the Undoability of Involvement." *European Journal of Social Theory* 12 (1): 117–33. https://doi.org/10.1177/1368431008099647.

Martin, Deborah L., Grace Gershuny, and Jerry Minnich, eds. 1992. *The Rodale Book of Composting*. New rev. ed. Rodale Press.

Martin, Douglas. 1999. "Rikers Island Learns to Recycle; Behind Razor Wire, Gardens and a Composting Plant." *The New York Times*, January 19.

Martinez, Miranda J. 2010. *Power at the Roots: Gentrification, Community Gardens, and the Puerto Ricans of the Lower East Side*. Lexington Books.

McCarthy, James E. 2003. "Interstate Waste Transport: Legislative Issues." RS20106. Congressional Research Service. https://crsreports.congress.gov/product/pdf/RS/RS20106/2.

McClintock, Nathan. 2014. "Radical, Reformist, and Garden-Variety Neoliberal: Coming to Terms with Urban Agriculture's Contradictions." *Local Environment* 19 (2): 147–71. https://doi.org/10.1080/13549839.2012.752797.

McKinley, Ira, and Bhawin Suchak, irs.. 2013. *The Throwaways*. Video. Third World Newsreel.

McLean, Katherine. 2011. "The Biopolitics of Needle Exchange in the United States." *Critical Public Health* 21 (1): 71–79. https://doi.org/10.1080/09581591003653124.

McNeur, Catherine. 2014. *Taming Manhattan: Environmental Battles in the Antebellum City*. Harvard University Press.

Melillo, Edward D. 2012. "The First Green Revolution: Debt Peonage and the Making of the Nitrogen Fertilizer Trade, 1840—1930." *The American Historical Review* 117 (4): 1028–60.

Melosi, Martin V. 2005. *Garbage in the Cities: Refuse, Reform, and the Environment*. Rev. ed. History of the Urban Environment. University of Pittsburgh Press.

———. 2008. *The Sanitary City: Environmental Services in Urban America from Colonial Times to the Present*. Abridged ed. History of the Urban Environment. University of Pittsburgh Press.

———. 2016. "Fresh Kills: The Making and Unmaking of a Wastescape." *RCC Perspectives*, 1:59–66.

Mencel, Aleks. 2014. "Canners versus the City—The Fight over Your Empties." *Brooklyn Based* (blog), April 24. https://brooklynbased.com/2014/04/24/canners-versus-the-city-the-fight-for-your-recycling-bin/.

Mercado, Angely. 2020. "What Does Pollution Have to Do with Police

Violence?" *Grist*, June 19. https://grist.org/justice/pollution-police-george-floyd-eric-garner-covid/.

Metcalfe, Alan, Mark Riley, Stewart Barr, Terry Tudor, Guy Robinson, and Steve Guilbert. 2012. "Food Waste Bins: Bridging Infrastructures and Practices." *The Sociological Review* 60 (December): 135–55. https://doi.org/10.1111/1467-954X.12042.

Meynell, Peter-John. 1976. *Methane: Planning a Digester*. C.T.T. Series; No. 2. Dorchester [Eng.]: Prism Press.

Millner, Patricia. 2020. "Reviewing Safe Procedures for Composting Feedstocks with Potential Coronavirus Contamination." Webinar presented at the Composting in the Time of COVID-19, April 1. https://cdn.ymaws.com/www.compostingcouncil.org/resource/resmgr/research_and_edu/uscc_sars-cov-2_covid19_webi.pdf.

Miraftab, Faranak. 2004. "Neoliberalism and Casualization of Public Sector Services: The Case of Waste Collection Services in Cape Town, South Africa." *International Journal of Urban and Regional Research* 28 (4): 874–92. https://doi.org/10.1111/j.0309-1317.2004.00557.x.

Mock, Brentin. 2014. "Think People of Color Don't Care about the Environment? Think Again." *Grist* (blog), July 29. https://grist.org/climate-energy/think-people-of-color-dont-care-about-the-environment-think-again/.

Mohai, Paul. 2003. "Dispelling Old Myths: African American Concern for the Environment." *Environment: Science and Policy for Sustainable Development* 45 (5): 10–26. https://doi.org/10.1080/00139150309604546.

Monani, Salma. 2009. "A Conversation with Julian Agyeman: Just Sustainability and Ecopedagogy." *Green Theory and Praxis: The Journal of Ecopedagogy* 5 (1): 51–68.

Montgomery, Jeff. 2014. "Odors Persist from Peninsula Compost Yard." *Delaware Online*, April 21. https://www.delawareonline.com/story/news/local/2014/04/19/odors-persist-peninsula-compost-yard/7918951/

Moran, Greta. 2021. "As Food Scrap Collection Peaks, NYC Community Compost Sites Face Evictions." *Civil Eats*, March 3. https://civileats.com/2021/03/03/as-food-scrap-collection-peaks-nyc-community-compost-sites-face-evictions/.

Moss, Jeremiah. 2012. "Disney World on the Hudson." *The New York Times*, August 21.

Moss, Michael. 1991. "Racing to Move Sludge: City Set to Award No-Bid Contracts." *Newsday*, January 29.

Moulaert, Frank, and Erik Swyngedouw. 2013. "Socially Innovative Projects, Governance Dynamics and Urban Change, between State and Self Organization." In *Can Neighbourhoods Save the City? Community Development and Social Innovation*, edited by Frank Moulaert, Erik Swyngedouw, Flavia Martinelli, and Sara Gonzalez. Regions and Cities. Routledge.

Mukhija, Vinit, and Anastasia Loukaitou-Sideris, eds. 2014. *The Informal American City: Beyond Taco Trucks and Day Labor*. MIT Press.

Muñoz, José Esteban. 2009. *Cruising Utopia: The Then and There of Queer Futurity*. NYU Press.

Munro, Kirstin. 2021. "Overaccumulation, Crisis, and the Contradictions of Household Waste Sorting." *Capital & Class*, 46 (1): 115–131. https://doi.org/10.1177/03098168211029004.

Muoio Dunn, Danielle. 2018. "De Blasio's Lofty Organics Goal Stalled by Lack of Funding, Participation." *Politico PRO*, September 4. https://politi.co/2wEJAnN.

Muoio Dunn, Danielle. 2020. "Council, Residents Look to Salvage Organics Recycling amid Budget CutsP." *Politico*, May 4. https://www.politico.com/states/new-york/city-hall/story/2020/05/03/council-residents-look-to-salvage-organics-recycling-amid-budget-cuts-1281759.

Murphy, Jarrett. 2015. "Life Near a Landfill: The Towns and People Who End Up with NYC Trash." *City Limits*, May 22. https://citylimits.org/2015/05/22/life-near-a-landfill-the-towns-and-people-who-end-up-with-nycs-trash/.

Nagle, Robin. 2013. *Picking Up: On the Streets and Behind the Trucks with the Sanitation Workers of New York City*. Macmillan.

National Grid. 2015. "Newtown Creek Renewable Natural Gas Proof of Concept Demonstration Project." Presented at NYC Organics is Expanding into CB1, Brooklyn, NY, June 3.

National Institute for Occupational Safety and Health. 2012. "NIOSH Fact Sheet: Solid Waste Industry." 2012–140. Department of Health and Human Services, National Institute for Occupational Safety and Health. https://www.cdc.gov/niosh/docs/2012-140/pdfs/2012-140.pdf.

Navarro, Mireya. 2013. "Bloomberg Plan Aims to Require Food Composting." *The New York Times*, June 16.

Nelson, Alondra. 2011. *Body and Soul: The Black Panther Party and the Fight against Medical Discrimination*. University of Minnesota Press.

Neuman, William. 2018. "New York's Push to End Inequality Extends to Garbage." *The New York Times*, August 16.

Newman, David M. 2016. "Dirty and Dangerous: Worker Safety and Health in New York City's Scofflaw Commercial Waste Industry." New York Committee for Occupational Safety and Health. https://nycosh.org/wp-content/uploads/2016/05/DirtyDangerous_FullReport_Final.pdf.

Newman, Kathe, and Elvin K. Wyly. 2006. "The Right to Stay Put, Revisited: Gentrification and Resistance to Displacement in New York City." *Urban Studies* 43 (1): 23–57. https://doi.org/10.1080/00420980500388710.

Nierenberg, Amelia. 2020. "Composting Has Been Scrapped. These New Yorkers Picked Up the Slack." *The New York Times*, August 9.

Nilsson, Anton K., and Tanay Warerkar. 2015. "City Insists Organics Program

Will Decrease Pollution Despite More Trucks." *Greenpoint Gazette*, June 11. http://greenpointnews.com/2015/06/11/city-insists-organics-program-will-decrease-pollution-despite-more-trucks/.

Nir, Sarah Maslin. 2016. "Community Gardens Imperiled by New York's Affordable Housing Plans." *The New York Times*, January 23.

Nolan, John. 1990. "Procter & Gamble Announces New Diapers, Composting Commitment Graphic." *AP News*, October 9.

Nonko, Emily. 2016. "Invasive Species: Developers See Green in Community Gardens." *Observer* (blog), *AP News*. January 28. https://observer.com/2016/01/invasive-species-developers-see-green-in-community-gardens/.

Norris-Waldt, Linda. 2018. "Silver Anniversary USCC: 25 Years and Growing." US Composting Council. https://www.compostingcouncil.org/page/silveranniversary.

Nye, David E. 1994. *American Technological Sublime*. MIT Press.

Obach, Brian K. 2016. *Organic Struggle: The Movement for Sustainable Agriculture in the United States*. MIT Press.

Offenhartz, Jake. 2022. "NYC Tries Again on Curbside Compost Collection, Minus the 'Psychodrama.'" *Gothamist* (blog), August 8. https://gothamist.com/news/nyc-tries-again-on-curbside-compost-collection-minus-the-psychodrama.

Okvat, Heather A., and Alex J. Zautra. 2014. "Sowing Seeds of Resilience: Community Gardening in a Post-Disaster Context." In *Greening in the Red Zone: Disaster, Resilience and Community Greening*, edited by Keith G. Tidball and Marianne E Krasny, 73–90. Springer Netherlands. https://doi.org/10.1007/978-90-481-9947-1_5.

Olds, Jerome. 2009. "A Statement of Purpose." *BioCycle*, April.

O'Leary, Philip R., and Patrick W Walsh. 1995. "Decision Makers' Guide to Solid Waste Management." Overviews & Factsheets. US Environmental Protection Agency, Office of Solid Waste. http://infohouse.p2ric.org/ref/03/02021/02021.pdf.

Orr, Steve. 2014. "Lawsuit Claims Organic Waste Meant for Composting Was Re-Routed to Upstate Landfill." *Democrat & Chronicle*, January 2.

Osman, Suleiman. 2012. *The Invention of Brownstone Brooklyn: Gentrification and the Search for Authenticity in Postwar New York*. Oxford University Press.

Otterbein, Holly. 2015. "Philly's Hottest New Bar Becomes Gentrification Battleground." *Philadelphia Magazine* (blog), September 4. http://www.phillymag.com/citified/2015/09/04/le-bok-fin-yelp-review/.

Ottinger, Gwen. 2013. *Refining Expertise: How Responsible Engineers Subvert Environmental Justice Challenges*. NYU Press.

Palmer, Paul. 2005. "Reusing Function, Not Materials." *Waste News* 10 (28): 9.

Parker, Julia Dawn, and Maureen H. McDonough. 1999. "Environmentalism of African Americans: An Analysis of the Subculture and Barriers

Theories." *Environment and Behavior* 31 (2): 155–77. https://doi.org/10.1177/00139169921972047.

Paull, David. 2019. "The Composting Culture." *BioCycle*, January.

Pellow, David N. 2016. "Toward a Critical Environmental Justice Studies: Black Lives Matter as an Environmental Justice Challenge." *Du Bois Review: Social Science Research on Race* 13 (2): 221–36. https://doi.org/10.1017/S1742058X1600014X.

Pellow, David N., Allan Schnaiberg, and Adam S Weinberg. 2000. "Putting the Ecological Modernization Thesis to the Test: The Promises and Performance of Urban Recycling." In *Ecological Modernizaton around the World: Perspectives and Critical Debates*, edited by Arthur P.J. Mol, David A. Sonnenfeld, 109–37. Frank Cass & Co.

Pellow, David Naguib. 2003. *Garbage Wars: The Struggle for Environmental Justice in Chicago*. MIT Press.

Penniman, Leah, and Karen Washington. 2018. *Farming While Black: Soul Fire Farm's Practical Guide to Liberation on the Land*. Chelsea Green Publishing.

Pinto, Maya, and Kristi Barnes. 2013. "Transform Don't Trash NYC: How to Increase Good Jobs, Recycling, and Justice in the Commercial Waste Industry." Alliance for a Greater New York.

Pires, Alda F. A., Patricia D. Millner, Jerome Baron, and Michele T. Jay-Russell. 2018. "Assessment of Current Practices of Organic Farmers Regarding Biological Soil Amendments of Animal Origin in a Multi-Regional U.S. Study." *Food Protection Trends* 38 (5): 347–62.

Platt, Brenda, Bobby Bell, and Cameron Harsh. 2013. "Pay Dirt: Composting in Maryland to Reduce Waste, Create Jobs, & Protect the Bay." Institute for Local Self-Reliance. https://cdn.ilsr.org/wp-content/uploads/2013/05/Pay-Dirt-Report.pdf.

Platt, Brenda, Nora Goldstein, Craig Coker, and Sally Brown. 2014. "State of Composting in the U.S." *BioCycle*, July.

Platt, Brenda, James McSweeny, and Jenn Davis. 2014. "Local Fertility: A Guide to Community Composting." Hardwick, VT: Highfields Center for Composting and the Institute for Local Self-Reliance. https://ilsr.org/wp-content/uploads/2014/07/growing-local-fertility.pdf.

Pollack, Shoshana, and Amy Rossiter. 2010. "Neoliberalism and the Entrepreneurial Subject: Implications for Feminism and Social Work." *Canadian Social Work Review / Revue Canadienne de Service Social* 27 (2): 155–69.

Potter, Cher, D. K. Osseo Asare, and Mugendi K. M'Rithaa. 2019. "Crafting Spaces between Design and Futures: The Case of the Agbogbloshie Makerspace Platform." *Journal of Futures Studies* 23 (3): 39–56. https://doi.org/10.6531/JFS.201903_23(3).0004.

Pulido, Laura. 2018. "Racism and the Anthropocene." In *Future Remains: A*

Cabinet of Curiosities for the Anthropocene, edited by Gregg Mitman, Marco Armiero, and Robert Emmett, 116–28. University of Chicago Press. https://doi.org/10.7208/9780226508825-014.

Pulido, Laura, and Juan De Lara. 2018. "Reimagining 'Justice' in Environmental Justice: Radical Ecologies, Decolonial Thought, and the Black Radical Tradition." *Environment and Planning E: Nature and Space* 1 (1–2): 76–98. https://doi.org/10.1177/2514848618770363.

Quastel, Noah. 2009. "Political Ecologies of Gentrification." *Urban Geography* 30 (7): 694–725. https://doi.org/10.2747/0272-3638.30.7.694.

Raffles, Hugh. 2011. "What to Do about Invasive Species." *The New York Times*, April 9.

Raloff, Janet. 1993. "Cleaning up Compost; Municipal Waste Managers See Hot Prospects in Rot." *Science News*, January 23.

Relis, Paul, and Howard Levenson. 1997. "Using Urban Organics in Agriculture." *BioCycle*, April.

Reno, Joshua O. 2016. *Waste Away: Working and Living with a North American Landfill.* University of California Press.

Reuter, Peter. 1993. "The Cartage Industry in New York." *Crime and Justice* 18:149–201.

Rheinberger, Hans-Jörg. 2012. "Experimental Systems: Difference, Graphematicity, Conjuncture." *Intellectual Birdhouse: Artistic Practice as Research*, edited by Florian Dombois, Ute Meta Bauer, Claudia Mareis, Michael Schwab, 89–99. Koenig Books.

Robinson, Solon. 1865. *Facts for Farmers; Also for the Family Circle.* A. J. Johnson; http://archive.org/details/factsforfarmersoorobi.

Rogers, Heather. 2015. "How Michael Bloomberg Greenwashed New York City." *Tablet*, January 5. https://www.tabletmag.com/sections/news/articles/planyc-bloomberg.

Rosan, Christina D. 2012. "Can PlaNYC Make New York City 'Greener and Greater' for Everyone? Sustainability Planning and the Promise of Environmental Justice." *Local Environment* 17 (9): 959–76. https://doi.org/10.1080/13549839.2011.627322.

Roudman, Sam. 2013. "Bank of America's Toxic Tower." *New Republic*, July 28. https://newrepublic.com/article/113942/bank-america-tower-and-leed-ratings-racket.

Roy, Ananya. 2005. "Urban Informality: Toward an Epistemology of Planning." *Journal of the American Planning Association* 71 (2): 147–58. https://doi.org/10.1080/01944360508976689.

———. 2009. "Why India Cannot Plan Its Cities: Informality, Insurgence and the Idiom of Urbanization." *Planning Theory* 8 (1): 76–87. https://doi.org/10.1177/1473095208099299.

Roy, Ananya, and Nezar AlSayyad. 2004. *Urban Informality: Transnational Perspectives from the Middle East, Latin America, and South Asia.* Lexington Books.

Royte, Elizabeth. 2006. *Garbage Land: On the Secret Trail of Trash*. Back Bay Books.
Rubinstein, Dana. 2023. "New York City Residents Will Soon Have to Compost Their Food Scraps." *The New York Times*, June 8.
Rubinstein, Dana, and Jeffery C. Mays. 2020. "Nearly $1 Billion Is Shifted from Police in Budget That Pleases No One." *The New York Times*, June 30.
R. W. Beck, Inc. 2010. "Hunts Point Anaerobic Digestion Feasibility Study." New York City Economic Development Corporation. http://www.nycedc.com/sites/default/files/filemanager/Projects/Hunts_Point_Peninsula/HuntsPointAnaerobicDigestionFeasibilityStudy.pdf.
Sandilands, Catriona. 2002. "Lesbian Separatist Communities and the Experience of Nature: Toward a Queer Ecology." *Organization & Environment* 15 (2): 131–63. https://doi.org/10.1177/10826602015002002.
Save Our Compost Coalition. 2020. "Save Our Compost Statement July 1, 2020." July 1. https://docs.google.com/document/d/e/2PACX-1vRIkguz-WM3MUTrFbt_AMpFWm3S1DwWrjHuNbIchDTp1gx9-5jO5F6cfB2AfuqlZgG_4Zf3NVedvMya/pub.
SCDHEC, Office of Solid Waste Reduction & Recycling. 2018. "Composting: A Guide for South Carolina Schools." https://scdhec.gov/sites/default/files/Library/OR-1520.pdf.
Schaffer, Guy. 2014. "Camp as a Politics of Waste." In *The Discard Studies Compendium*, edited by Max Liboiron, Michele Acuto, and Robin Nagle. https://discardstudies.com/discard-studies-compendium/#Camp.
Schmelzkopf, Karen. 1995. "Urban Community Gardens as Contested Space." *Geographical Review* 85 (3): 364–81. https://doi.org/10.2307/215279.
Schneider, Daniel. 2011. *Hybrid Nature: Sewage Treatment and the Contradictions of the Industrial Ecosystem*. MIT Press.
Schouten, John W., Diane M. Martin, and Jack S. Tillotson. 2014. "Curbside Cartographies in an Urban Food-Waste Composting Program." In *Waste Management and Sustainable Consumption: Reflections on Consumer Waste*, edited by Karin M. Ekström, 102–14. Routledge.
Schrader, Astrid. 2010. "Responding to *Pfiesteria Piscicida* (the Fish Killer): Phantomatic Ontologies, Indeterminacy, and Responsibility in Toxic Microbiology." *Social Studies of Science* 40 (2): 275–306. https://doi.org/10.1177/0306312709344902.
Schulman, Sarah. 2012. *The Gentrification of the Mind: Witness to a Lost Imagination*. University of California Press.
Seattle. 2019. "Recycling and Composting Benefits." https://www.seattle.gov/utilities/your-services/collection-and-disposal/multi-family-properties/recycling-and-composting-benefits.
Sedgwick, Eve Kosofsky. 2003. *Touching Feeling: Affect, Pedagogy, Performativity*. Duke University Press.
Sengupta, Somini. 1996. "Sewage Plant: No vs. Maybe." *The New York Times*, September 22.

Seyfang, Gill, and Adrian Smith. 2007. "Grassroots Innovations for Sustainable Development: Towards a New Research and Policy Agenda." *Environmental Politics* 16 (4): 584–603. https://doi.org/10.1080/09644010701419121.

Shapin, Steven, and Simon Schaffer. 1985. *Leviathan and the Air-Pump: Hobbes, Boyle, and the Experimental Life*. Princeton University Press.

Shepard, Peggy, Stephanie Tyree, and Cecil Corbin-Mark. 2008. "PlaNYC: EJ Group Takes the Inside Track to Advocate Sustainability." *Who Owns Our Cities?* 15 (1): 70–72.

Shimpo, Naomi, Andreas Wesener, and Wendy McWilliam. 2019. "How Community Gardens May Contribute to Community Resilience Following an Earthquake." *Urban Forestry & Urban Greening* 38 (February): 124–32. https://doi.org/10.1016/j.ufug.2018.12.002.

Shove, Elizabeth, and Gordon Walker. 2010. "Governing Transitions in the Sustainability of Everyday Life." *Research Policy* 39 (4): 471–76. https://doi.org/10.1016/j.respol.2010.01.019.

Showstack, Randy. 1997. "Fears of Asthma Fuel Protests over a Waste Incinerator." *The New York Times*, May 11.

Skumatz, Lisa A. 2008. "Pay as You Throw in the US: Implementation, Impacts, and Experience." *Waste Management* 28 (12): 2778–85. https://doi.org/10.1016/j.wasman.2008.03.033.

Slocum, Rachel. 2006. "Anti-Racist Practice and the Work of Community Food Organizations." *Antipode* 38 (2): 327–49. https://doi.org/10.1111/j.1467-8330.2006.00582.x.

Smil, Vaclav. 2001. *Enriching the Earth: Fritz Haber, Carl Bosch, and the Transformation of World Food Production*. MIT Press.

smiley, sam, dir. 2014. *The Queen of Kudzu*. Video. https://vimeo.com/83998526.

Smith, Adrian. 2012. "Civil Society in Sustainable Energy Transitions: Adrian Smith." In *Governing the Energy Transition: Reality, Illusion or Necessity?*, edited by Geert Verbong and Derk Loorbach, 180–202. Routledge.

Smith, Adrian, Adrian Ely. 2015. "Green Transformations from Below? The Politics of Grassroots Innovation." In *The Politics of Green Transformations*, edited by Ian Scoones, Melissa Leach, and Peter Newell, 102–18. Routledge.

Smith, Dorothy E. 2005. *Institutional Ethnography: A Sociology for People*. AltaMira Press.

Smith, Gillian. 2020. "Montgomery County Unveils New Food Scraps Recycling Program." *Bethesda-Chevy Chase Patch*, April 22. https://patch.com/maryland/bethesda-chevychase/montgomery-county-unveils-new-food-scraps-recycling-program.

Smith, J. Waldo. 1924. "The Beccari Process of Garbage Disposal." *American Journal of Public Health* 14 (2): 119–21. https://doi.org/10.2105/AJPH.14.2.119.

Specter, Michael. 1992. "Among the Earth Baby Set, Disposable Diapers Are Back." *The New York Times*, October 23.

St. Hillaire, Dior. 2020. "Compost Capitalism Dior St Hillaire." Zero Waste Conference 2020, October 28. Presentation recording, 1:23:18. https://www.youtube.com/watch?v=_vZmyLtCdnU.

Star, Susan Leigh. 1999. "The Ethnography of Infrastructure." *American Behavioral Scientist* 43 (3): 377–91. https://doi.org/10.1177/00027649921955326.

Stein, Perry. 2017. "D.C. Wants to Give Every Resident a Bin for Composting and Offer Curbside Pickup in Five Years." *Washington Post*, August 7.

Stephens, Jennie C. 2019. "Energy Democracy: Redistributing Power to the People through Renewable Transformation." *Environment: Science and Policy for Sustainable Development* 61 (2): 4–13. https://doi.org/10.1080/00139157.2019.1564212.

Sternberg, Carolina Ana. 2013. "From 'Cartoneros' to 'Recolectores Urbanos.' The Changing Rhetoric and Urban Waste Management Policies in Neoliberal Buenos Aires." *Geoforum* 48 (August): 187–95. https://doi.org/10.1016/j.geoforum.2013.04.029.

Stevens, Elizabeth Lesly. 2011. "Picking Up the City's Garbage Is a Sweet Deal, and a Monopoly." *The New York Times*, June 9.

Stewart, Kathleen. 2007. *Ordinary Affects*. Duke University Press.

Stewart, Kristyn. 2015. "Scapegoating Le Bok Fin Missed the Real Problem for Philly Schools." *WHYY*, September 17. https://whyy.org/articles/scapegoating-le-bok-fin-missed-the-real-problem-for-philly-schools/.

Storey, Phil. 2005. "Casting a Long Shadow." *Green@work*, Summer. http://www.greenatworkmag.com/gwsubaccess/05summer/cover.html.

Strasser, Susan. 1999. *Waste and Want: A Social History of Trash*. Macmillan.

Streeter, Virginia, and Brenda Platt. 2017. "Residential Food Waste Collection Access in the U.S." The J. G. Press, Inc.

Stringer, Scott M. 2018. "Unsafe Sanitation: An Analysis of the Commercial Waste Industry's Safety Report." Office of the New York City Comptroller, Bureau of Budget and Bureau of Policy and Research. https://comptroller.nyc.gov/wp-content/uploads/documents/Commercial-Waste.pdf.

Swyngedouw, Erik. 2009. "The Antinomies of the Postpolitical City: In Search of a Democratic Politics of Environmental Production." *International Journal of Urban and Regional Research* 33 (3): 601–20. https://doi.org/10.1111/j.1468-2427.2009.00859.x.

Sze, Julie. 2006. *Noxious New York: The Racial Politics of Urban Health and Environmental Justice*. MIT Press.

TallBear, Kim. 2014. "Standing with and Speaking as Faith: A Feminist-Indigenous Approach to Inquiry." *Journal of Research Practice* 10 (2).

Tavory, Iddo, and Nina Eliasoph. 2013. "Coordinating Futures: Toward a Theory of Anticipation." *American Journal of Sociology* 118 (4): 908–42.

Taylor, Dorceta E. 2014. "The State of Diversity in Environmental Organizations." Green 2.0.

Telhan, Orkan, Yasmin B. Kafai, Richard Lee Davis, Kfai Steele, and Barrie M. Adleberg. 2014. "Connected Messages: A Maker Approach to Interactive Community Murals with Youth." In *Proceedings of the 2014 Conference on Interaction Design and Children*, 193–96. IDC '14. Association for Computing Machinery. https://doi.org/10.1145/2593968.2610450.

Tierney, John. 1996. "Recycling Is Garbage." *The New York Times Magazine*, June 30.

Tilth Alliance. 2019. "What Is Compost?" March 23. https://tilthalliance.org/resources/what-is-compost.

Tonge, Peter. 1987. "Seattle's 'Master Composters': They Spread the Good Word about Easy Home Composting." *Christian Science Monitor*, December 8. http://www.csmonitor.com/1987/1208/hcompo.html.

Torres-Velez, Victor M. 2011. "Puerto Ricans and the Green Jobs Gap in New York City." *Centro Journal* 23 (2): 95–113.

Travers, Eugene. 2010. "Douglas Durst Discusses Current Projects and Sustainable Development in the City." *CityLand* (blog), December 15. http://www.citylandnyc.org/douglas-durst-discusses-current-projects-and-sustainable-development-in-the-city/.

Treat, Sharon Anglin. 2021. "Forever Chemicals and Agriculture Case Study: Maine Accelerates Across-the-Board Action to Address PFAS Chemicals Harming Farmers and Rural Communities." Institute for Agriculture and Trade Policy. https://www.iatp.org/documents/forever-chemicals-and-agriculture-case-study.

Tsing, Anna Lowenhaupt. 2015. *The Mushroom at the End of the World: On the Possibility of Life in Capitalist Ruins*. Princeton University Press.

US Department of Agriculture. 2019. "Fertilizer Use and Price." https://www.ers.usda.gov/data-products/fertilizer-use-and-price/.

van der Ploeg, R. R., W. Böhm, and M. B. Kirkham. 1999. "On the Origin of the Theory of Mineral Nutrition of Plants and the Law of the Minimum." *Soil Science Society of America Journal* 63 (5): 1055–62. https://doi.org/10.2136/sssaj1999.6351055x.

Varda, Agnès, dir. 2000. *The Gleaners & I*. Video. Zeitgeist Films.

Vasintjan, Aaron. 2015. "Food Banks: Canned Justice or Fuel for Activism?" *Discard Studies* (blog), July 7. http://discardstudies.com/2015/07/07/food-banks-canned-justice-or-fuel-for-activism/.

von Wirth, Timo, Lea Fuenfschilling, Niki Frantzeskaki, and Lars Coenen. 2019. "Impacts of Urban Living Labs on Sustainability Transitions: Mechanisms and Strategies for Systemic Change through Experimentation." *European Planning Studies* 27 (2): 229–57. https://doi.org/10.1080/09654313.2018.1504895.

Wajcman, Judy. 1991. *Feminism Confronts Technology*. Polity Press.

Walls, Margaret. 2006. "Extended Producer Responsibility and Product Design: Economic Theory and Selected Case Studies." SSRN Scholarly Paper. https://doi.org/10.2139/ssrn.901661.

Walters, Joanna. 2018. "'Ick': Rats, Roaches and Rank Smells Dampen New York's Compost Program." *The Guardian*, October 7. https://www.theguardian.com/us-news/2018/oct/07/ick-rats-roaches-rank-smells-dampen-nyc-composting-program.

Wen, Qi-Xiao. 1984. "Utilization of Organic Materials in Rice Production in China." In *Organic Matter and Rice*, edited by Stephen Banta and Corazon V. Mendoza, 45–56. International Rice Research Institute.

Whitford, Emma. 2015. "This Wes Andersony Crochet Graffiti Is Causing Extreme Bushwick Gentrification Anxiety." *Gothamist* (blog), September 18. http://gothamist.com/2015/09/18/bushwick_crochet_gentrification.php.

williams, hassan, dir. 2015. *The "Trash." T-Shirt Project*. April 6. YouTube video, 3:41. https://www.youtube.com/ (no longer available).

Winner, Langdon. 1980. "Do Artifacts Have Politics?" *Daedalus* 109 (1): 121–36.

Wolch, Jennifer R., Jason Byrne, and Joshua P. Newell. 2014. "Urban Green Space, Public Health, and Environmental Justice: The Challenge of Making Cities 'Just Green Enough.'" *Landscape and Urban Planning* 125 (May): 234–44. https://doi.org/10.1016/j.landurbplan.2014.01.017.

Wood, Justin. 2019. "Trashing New York's Neighborhoods." Transform Don't Trash NYC. http://transformdonttrashnyc.org/wp-content/uploads/2019/10/trashing-new-yorks-neighborhoods.pdf.

Woods, William I., and Joseph M. McCann. 1999. "The Anthropogenic Origin and Persistence of Amazonian Dark Earths." *Yearbook. Conference of Latin Americanist Geographers* 25:7–14.

Wright, Erik Olin. 2010. *Envisioning Real Utopias*. Verso.

Wynne, Brian. 1992. "Misunderstood Misunderstanding: Social Identities and Public Uptake of Science." *Public Understanding of Science* 1 (3): 281–304. https://doi.org/10.1088/0963-6625/1/3/004.

Yardley, Jim. 1998. "Garbage in . . . and in . . . and in; Greenpoint Residents Unite to Fight Influx of Trash." *The New York Times*, April 18.

Yepsen, Rhodes. 2015. "Residential Food Waste Collection in the U.S." *BioCycle*, January.

Zimnicki, Tom. 2019. "Agriculture, Compost and Water Quality Intersections." *BioCycle*, January.

Zitcer, Andrew. 2015. "Food Co-Ops and the Paradox of Exclusivity." *Antipode* 47 (3): 812–28. https://doi.org/10.1111/anti.12129.

Zukin, Sharon, and Laura Braslow. 2011 "The Life Cycle of New York's Creative Districts: Reflections on the Unanticipated Consequences of Unplanned Cultural Zones." *City, Culture and Society* 2 (January): 131–40. https://doi.org/10.1016/j.ccs.2011.06.003.

Index

Note: References in italic refer to figures. References followed by "n" refer to notes.

"oX30," 2, 58, 78, 176
596 Acres, 104, 117, 117n1

abolition ecology, 36, 37, 120
abstract utopia, 16, 73, 114–15, 174. *See also* cruel optimism
actinomycetes, 129
Action Carting company, 81
Adams, Eric, 176
Added Value Compost, at Red Hook Farm, 88, 172
aeration, 39, 90, 95, 129. *See also* static aeration
aerobic decomposition, 39, 129–30. *See also* aeration
affordable housing, 15, 123
Agyeman, Julian, 120. *See also* just sustainability
Alkon, Alison, 121, 154
Alliance for a Greater New York, 69
anaerobic composting, 38, 96
anaerobic digestion (AD), 40n6, 45, 65–67
Anderson, Bridget, 65–66, 100
Anguelovski, Isabelle, 120, 127
anticipatory illumination, 145
Bachman, Emily, 69–70
backyard composting pilots, 52–53, 94
Baltimore Compost Collective, 121
Bambara, Toni Cade, 17
Bayrer, Charlie, 79–82, 85–87, 97, 99, 104, 111–12. *See also* Earth Matter
Bedminster-Marlborough facility, 64
Bennhold, Katrin, 54
Bergmann, Frithjof, 147
Berlant, Lauren, 174
Big Reuse, 22, 77, 101–2, 107, 170–72; conception of community, 109–10; site, 88–92, 104
BioCycle magazine, 40–43, 45–46, 110, 173
biogas, 65–66, 145. *See also* methane
biosolids, 40, 40n6, 56. *See also* sewage sludge
BK ROT, 17–18, 22–25, 117–59: and Brotherhood Sister Sol, 138, 149, 169; commercial collection, 142, 154–56; compensating waste labor, 92, 118, 147, 152–57; data collection, 103–4; and environmental justice, 118–24, 130–31, 171; Hilary Allison's comic, *137*; nonprofit structure, 118, 151–52;

BK ROT (*continued*)
pandemic response, 156–57; residential collection, 117, 142–43, 154, 156; and the #SaveOurCompost campaign, 106; vision for waste, 110–11, 142, 144–45, 153–59, 171; youth involvement in, 117, 142, 146–47, 150. *See also* Know Waste Lands; Nurse, Sandy; Peperone, Renée
Black Lives Matter, 123
Bloch, Ernst, 2, 16, 16n6. *See also* abstract utopia; concrete utopia
Bloomberg, Michael, 53–56, 65n21
Boeva, Yana, 99
Boggs, Grace Lee, 147
bokashi method, 96–97
Botanical Gardens, 52–53, 77, 95, 97, 132
boundary organization, 76
Bowker, Geoffrey, 21
Brewer, Gale, 69
Broad City (sitcom), 4
Brooklyn Botanic Garden, 52, 95, 97, 132
Brooklyn Transfer, 21, 61
Brotherhood Sister Sol, 93, 138, 149, 169
Brown, Sally, 173
The Brown Bike Girl, 147
brown bins. *See* curbside bins
Bruce, Louise, 76–79, 82n4, 92, 93n9, 101–3, 108–9, 113
Bulkeley, Harriet, 166
Bullard, Robert, 130
Bureau of Recycling and Sustainability, at DSNY, 21, 24, 76, 78
Bureau of Waste Prevention, Reuse and Recycling (BWPRR), at DSNY, 78–79
Bushwick: gentrification of, 118n3, 122–28, 140; long-term residents, 140–41; and municipal organics, 154; and waste infrastructure, 120, 130–31
Bushwick Abbey, 118
Business Integrity Commission (BIC), 15, 92–93, 113, 109, 114, 163–64, 172
busy-ness, 12, 18, 73, 114, 164, 167, 177. *See also* MacBride, Samantha
Byrne, Jason, 126
capitalist realism, 10
Carbone v. Clarkstown, 41n7
Carver, George Washington, 36–39
Centralized Organic Recycling equipment (CORe), 65, 67
CERO Cooperative, in Dorchester, 120–21
Cetiliztli Nauhcampa, *128*
Chancey, Bahij, 51
Checker, Melissa, 54
Christiansen, Kendall, 67
Cipollone, Melissa, 101
Citizens Budget Commission, of NYC, 67, 69, 71
Citywide Recycling Advisory Board (CRAB), 51–52
civic infrastructuring, 114
colonialism, 3, 5, 35–36, 120, 131, 165
commercial haulers: legal landscape, 41n7; market for, 13, 72, 163; and microhaulers, 93, 171; and NYC's waste, 4, 21, 62; in the waste regime, 6–7
Common Ground Compost, 22, 92, 106–7, 164
community composting: building community, 108–12; and the curbside organics collection pilots, 2–3, 8–9, 45, 56, 69–70, 74; during COVID-19 pandemic, 88, 99, 106, 156; definition of, 96, 101, 106, 109–11, 113; and volunteer labor, 14, 84–88, 90, 93, 103, 106, 113–14
Community Composting Council, 92–93, 95, 100–101, 106, 109
Community Composting Report, 22, 101, 104, 109, 167
community economies, 112, 158
community gardens/gardeners: and BK ROT, 117; and developers, 15; and food, 134–35; and gentrification, 124; housing community compost, 102, 105; and labor, 153; and master composters, 94; and microhaulers, 106, 171; and the NYC Compost Project, 76, 79–80, 82–83, 93, 101n12.

See also Compost for Brooklyn; El Garden; Know Waste Lands
community solar, 165, 177
compost (substance), 1, 10, 95, 143–45, 162; in agriculture, 34–38, 40–41, 163; for community gardens, 45, 80, 82; mixed MSW, 40–41; for remediation, 45, 102–3; selling, 41–42, 64, 92–93, 172; as vehicle for fertilizer, 39
Compost for Brooklyn (C4B), 22, 102–3, 108–9, 161, 168, 170
composting. *See* community composting; composting (metaphor); industrial composting; municipal composting
composting (metaphor), 3, 19–20, 74, 157–59, 161, 163, 173–78
compost pilots. *See* curbside organics collection pilots
Compost Promenade at Earth Matter, 97, *98*
Compost Science magazine. See *BioCycle* magazine
compost tumblers, 97, 99, 108, 148–49, 153
concrete utopia, 2, 16–17, 76–77, 115, 142, 145, 175
Cooperation Jackson, 165, 177
Covert, Matt, 105
COVID-19 pandemic: and BK ROT, 142–43, 156–57, 171; and community composting, 88, 99, 106–7; and municipal composting, 3, 7, 67–72, 107, 142–43, 172, 176
critical environmental justice, 5, 131
cruel optimism, 174–75
Cthulhucene, 19, 19n7
curbside bins, 44, 48, 58–60, 68
curbside organics collection pilots, NYC, 2–3, 15, 48–74; and BK ROT, 142, 154; and community compost, 76–77, 83, 103; and the multilevel perspective, 7–9; in national context, 29–31, 45; researching, 20–24; as utopia, 157, 175–76
Cvetkovich, Ann, 157, 168
Dano Process, 38
de Blasio, Bill, 49, 56, 58–59, 68, 70, 72, 78, 106
DeDominicis, Marisa, 79–82, 85–87, 97–98. *See also* Earth Matter
de Luco, Ana, 92, 135
Denver Compost Collective, 120
Department of Environmental Protection, NYC (DEP), 21, 65, 67
Department of Housing Preservation and Development, NYC, 15, 123
Department of Parks and Recreation, NYC, 15, 21, 61, 107, 171
Department of Sanitation, NYC (DSNY), 20–22; and BK ROT, 142, 154, 171–72; and community compost, 2–3, 8, 15, 75–116, 167, 171–72, 175; curbside organics collection, 7–8, 13, 29, 48–74, 157–58, 175; methodological hurdles, 23–24; waste stream, 3–4; workforce, 6. *See also* curbside organics collection pilots
diaper waste, 42–43
Dillon, Lindsey, 124
DiSalvo, Carl, 141
disgust, 18, 60, 77, 95, 128–30, 136, 162, 168. *See also* "ick factor"
Durst Organization, 57, 62–63
"dystopian events repackaged as feel-good story" genre, 9

Earth Matter, 22, 77, 79–82, 84–88, *86*, 90, 97–99, *98*, 101, 170–72
El Garden, 106, 118, 127, 130–31, 133–35, *148*, 149
Eliasoph, Nina, 12, 90, 139
emotions. *See* feelings
environmental injustice, 5–6, 18, 62, 120, 123, 131, 159
environmental justice: approach to compost, 22, 118, 120–21, 130–31; and community recycling, 135; and environmentalism, 36; movements, 6–8, 69; and municipal compost, 49, 69, 176; and PlaNYC, 53

experimental infrastructures: for composting, 1–3, 75, 112–13, 118, 159; in sustainability transitions, 161, 164–73; utopian, 2, 16–18, 175, 177

Farming Concrete, 104
feelings, 12–13, 17–18, 24–25, 130, 139, 168, 173–74. *See also* disgust; hope
fertilizers, 35–39, 55, 56n12
"Fight for $15" campaign, 22n10, 117
Fiore, Anthony, 66–67
Fisher, Mark, 10
food waste: abjection of, 136; under Bloomberg, 53–56; in households, 59–60, 90; as an issue, 13; rescuing, 21; as a resource, 92, 161–62; systems for recycling, 21, 29–31, 50–53
formality, 14–16, 100–106, 113, 165–66, 171–72
Fort Greene Compost Project, 80, 83, 111. *See also* Earth Matter
Foster, Ellen, 99
Fresh Kills Landfill, 6, 50, 52, 61

Garcia, Kathryn, 55, 68, 71, 78
gentrification, 54, 118n3, 120, 122–28, 133–34, 155
Gibson-Graham, J. K., 10, 112, 158, 169, 177
Giuliani, Rudy, 52
Givelove, 45
Goldstein, Jerome (Jerry), 40
Goldstein, Joshua, 15n4
Goldstein, Nora, 41
Golueke, Clarence, 39, 41, 72–73
"good green jobs," 22, 128, 146–47, 151n16, 170
Goodyear, Sarah, 125–26
Gordon, Avery, 17
Gould, Kenneth, 120, 125
grassroots compost. *See* community composting
grassroots innovation, 7–9, 14, 169
Grayson, Eric, 70–71
Green, Justin, 70
Greenberg, Miriam, 55–56
green economy, 121, 154, 158
GreenFeen OrganiX, 121, 158, 163
"green for all," 154
green gentrification, 54, 120, 155
Green Guerillas, 128n6
green infrastructure, 125
Green Thumb, 92–93, 138
GrowNYC, 77, 80–81, 84, 88, 107, 172
guano, 35–36
Gurley, Akai, 123n4
Guston, David, 76
Guthman, Julie, 134

habits, 4, 17, 57n14, 58–59, 73–74, 111, 157, 167–69
Hackworth, Jason, 123
Hamilton, Jennifer Mae, 19n8, 158
Haraway, Donna, 19, 158, 168
Hard Core Compost, in Santa Cruz, 120
Hawkins, Gay, 4, 73, 111, 168
Helena Building, 57, 63
Hersey, Mark, 36
Hess, David, 96, 146
Heynen, Nik, 37, 120, 131
The High Line, 125
hope: at BK ROT, 144–45; and busyness, 114; and composting, 173–77; in experimental infrastructures, 1, 3, 161, 168; and habit, 157, 175; methodological considerations, 26; and recycling, 30–31, 44; and utopia, 16–18, 145. *See also* abstract utopia; concrete utopia; cruel optimism
Howard, Albert and Gabrielle, 36–40

"ick factor," 51. *See also* disgust
incineration, 4, 33–34, 50, 52, 121
individualization, 11–13, 72, 160–61, 164–65, 167
Indore process, 36–37
industrial composting, 38–42, 45–46, 60–64, 96
informality, 1, 14–16, 32, 76, 100–107, 164–65, 171–72
infrastructure: assembling, 105–6; decentralized, 101; emotional experiences of, 18, 13, 130, 168; and environmental injustice, 5–6, 11,

17, 130–32, 162; ethnography of, 21; grassroots, 1–3, 9, 25, 165, 175; and households, 58–60, 113; invisibility of, 4–5; living with, 111–12, 131; poetics of, 5, 84; prefigurative, 165–66, 177. *See also* experimental infrastructures; formality; informality
in-sink disposal, 67
InSinkErator. *See* in-sink disposal
institutional ethnography, 23
Irving Farm, 63

Jacobs, Jane, 108, 138
Jacobson, Abbi, 4
Jalbert, Kirk, 105
James, Leititia, 54–55
Jaylor, 89–90, 89n7, *91*
Jensen, Casper Bruun, 18, 166
Johnson, Corey, 70
Johnson, James, 62
"just green enough" environmentalism, 126, 126n5
just sustainability, 9, 120, 130–31, 156–57

Keil, Roger, 16n5
Keller, Evelyn Fox, 98
Kenner, Alison, 114
Kenney, Martha, 168
Kinchy, Abby, 105
Know Waste Lands (KWL), 17; concrete pad at, 143–46; and gentrification, 126; murals at, 139–40; and native plants, 141n12; open house at, 127, *128*; origins of, 118, 118n3; signage, *119*; as waste infrastructure, 135, 138. *See also* BK ROT

labor. *See* waste labor
landfills, 3–5, 11, 34, 41–42, 46, 69
landfill utopianism, 19–20, 157, 173
Larkin, Brian, 4–5
Laundromat Project, 140, 140n10
leachate, 61, 132, 136
Levitas, Ruth, 16
Lewis, Tammy, 125
Liboiron, Max, 32n2
Local Law 19 of 1989, 50
Local Law 77 of 2013, 21, 57, 60
Local Organics Recovery Program, 77, 81
Loop Closing Compost, 169–70
Lorde, Audre, 18
Lower East Side Ecology Center (LESEC), 22, 75, 77, 80, 101, 107, 171, 172
Lyons, Jessica, 105

MacBride, Samantha, 11–12, 18, 30, 57n14, 73–74, 164. *See also* busy-ness
Mackinac Island, 29
Madanipour, Ali, 124
mandatory organics recycling, 70–72, 172, 176
Manela, Marguerite, 78, 79, 102, 109
Marres, Noortje, 170
Master Composter Certification Course (MCCC), 77, 81, 93–95, 97, 99, 113, 171
Master Composter Manual (MCM), 23, 87, 92–93, 95, 96–97, 103, 170
master composters, 94–95, 161
master narratives, 21
Mayor's Food Waste Challenge, 53–54
McEnroe, Erich, 62–64
McEnroe Compost, 46, 57, 62–64, 163
McNeur, Catherine, 32n3
Metcalfe, Alan, 59
methane, 40n6, 56, 65–66, 129
Michigan State University, 38–39
microhaulers, 13, 15, 22, 104, 106–7, 120, 163–64, 171–72
mixed waste composting, 39–40, 41, 43, 44n10
Morales, Alfonso, 105
Morales, Domingo, 82n4, 132–33, 138, 144, 152, 159
Morita, Atsuro, 18, 166
Morrison, Kendall, 80
multilevel perspective, 6–7, 14–15, 165
municipal composting, 2–3, 6–9, 29–31, 38–47. *See also* curbside organics collection pilots
municipal waste collection, 11, 32–34. *See also* municipal composting
Muñoz, José Esteban, 31, 115, 145, 159
Munro, Kristin, 71, 90n8

Nagle, Robin, 6, 89–90, 153
National Grid, 65–67
National Recycling Coalition, 43–44
Natural Resources Defense Council (NRDC), 52, 55
Neimanis, Astrida, 19n8, 158
neoliberal waste management, 9–17, 161–64; and BK ROT, 122, 155, 157–58; and community compost, 113–14; and the curbside organics collection pilot, 49, 54–56, 72, 76; depoliticized, 17, 31; and repurposing waste, 135–36. *See* waste: markets for
Newell, Joshua, 126
Newman, Kathe, 123
New Milford Farms, 64
Newtown Creek Wastewater Treatment Plant, 60, 66–67
New York City Compost Project, 3, 75–116; and the Community Composting Council, 100–101; conception of community, 109–12; creation of, 51, 77–78; and a "culture of composting," 20, 51, 76, 167, 169, 172; formalizing grassroots systems, 15, 79–84, 93, 100–106, 172; funding cuts, 68, 106–7; and neoliberal waste management, 113–14; relationships with waste, 167; structure of, 75–79. *See also* community composting; Master Composter Certification Course
New York City Department of Sanitation. *See* Department of Sanitation
NIMBYism, 130
Nurse, Sandy, 92, 99, 118, 127, 130–32, 139, 143n13, 147, 149–53, 169–71; approach to compost, 22, 110–11, 130–32, 153; approach to green jobs, 147, 150–51, 153; building KWL, 140, 143; as council member, 118, 176; and the origins of BK ROT, 117–18; on policing, 123; at the #SaveOurCompost Town Hall, 69. *See also* BK ROT

Office of Long-Term Planning and Sustainability, 53
OneNYC plan, 48, 53, 58, 78, 103
opt-in organics recycling, 49, 70–73
organics recycling, 29–47. *See also* curbside organics collection pilots; industrial composting; municipal composting
organic waste. *See* food waste
Oyewole, Tok Oluwaseyi, 69

pandemic. *See* COVID-19 pandemic
paranoid and reparative readings, 24–25, 76, 114
Parks Department. *See* Department of Parks and Recreation
Park Slope, 50, 57, 59
Pataki, George, 52
pay-as-you-throw (PAYT) policies, 11
Pedilla, Carlos, 130
Peninsula Compost Facility, 61–62, 64, 81, 162
Peperone, Renée, 22, 95, 99, 134, 135n8, 143, 146–53, 169–71; approach to compost, 22, 130–32; on gentrification, 125–27; and the origins of BK ROT, 117–18; relationship with youth workers, 139, 146, 150–52
permaculture, 56, 94, 132
Pineda, Ceci, 118, 142, 147, 153, 156–57, 159–60, 169
PlaNYC, 53–55
police violence, 123–24, 139
postpolitical sustainability, 10, 17
private waste industry. *See* commercial haulers; industrial composting
Procter & Gamble, 43
Pulido, Laura, 120, 122

Quinn, Christine, 81

Rabbit Roots, 142–43, 156
racial capitalism, 36, 38, 120, 158
racism, 5, 36, 121, 120, 123, 130–32
radical ecology, 36, 120, 122
"Recycle First" report (CRAB), 52, 52n4
recycling: and busy-ness, 19, 30, 74; community centers for, 42, 135; in households, 59, 71; and neoliberalism, 11; in New York City, 50–52, 52n4;

origins of, 33, 42–44; and subjectivity, 73, 111. *See also* organics recycling
Red Hook Farms, 22, 80, 88
Reyna, Diana, 57n13
Reynoso, Antonio, 68, 71
Rheinberger, Hans-Jorg, 166
Rikers Island Prison Complex, 52, 52n5
Rogers, Heather, 53
Rose, Deborah L., 56
Rosenshine, Laura, 92
Roy, Ananya, 14, 171

Sacks, Anna, 106
sacrifice zones, for managing wastes, 5
salvage accumulation, 14
sanitation workers, 6, 89–90. *See also* waste labor
Sapelo Island Cultural and Revitalization Society, 165, 177
#SaveOurCompost campaign, 49, 69, 72, 74, 100, 106
Schrader, Astrid, 168
science and technology studies, 3, 166
Sedgwick, Eve Kosofsky, 24. *See also* paranoid and reparative reading
sewage sludge, 39, 40n6, 56, 65, 66n22, 67
Seyfang, Gill, 8
Sheintoch, Debbie, 81, 83
skid-steer, 89, 90
Slocum, Rachel, 134
Smith, Adrian, 7–8, 164
Smith, Dorothy, 23
Smith, Neil, 123
social justice, 120–21, 132, 154, 160, 163
social movements: and grassroots infrastructures, 1; for hygiene, 32–34; and knowledge, 96; for organic agriculture, 37–38; for organics recycling, 46; against police violence, 123–24; for recycling, 44; and sustainability transitions, 7–9
soil fertility, 34–38, 40–42, 132
Solid Waste Act of 1965, 40
Solid Waste Advisory Board (SWAB), 51–52, 55
Solid Waste Composting Council, 43–44
Solid Waste Management Plan (2006), 65n21
Soul Fire Farm, 165, 177
source separation, 8, 33, 43–44, 47, 50–51, 58, 73
Star, Susan Leigh, 4, 21
Starrett City, Brooklyn, 50, 51
static aeration, 61, 89, 104
Stenn, Howard, 94
Sternberg, Carolina Ana, 15n4
Stewart, Kristyn, 126
St. Hillaire, Dior, 70, 158, 163, 178
Strasser, Susan, 32
Sure We Can, 135
sustainability transitions, 6–9, 121, 154, 159, 161, 164–73
Swyngedouw, Erik, 17
Sze, Julie, 124

TallBear, Kim, 25
Tavory, Iddo, 139
Thorburn, Vandra, 92, 96n11
toter tipper, 85, *86*, 89–90, 99
tipping fees, 11, 61, 63–64, 69, 92, 106, 153, 162–64
trichinosis, 33
Tsing, Anna, 14

University of California at Berkeley, 38
UPROSE, 53n8
USDA Soil Conservation Service, 40

Van Woert, Jim, 65–66
Veteran Compost, 121
Vokashi, 22, 92, 96n11, 106, 164
voluntary organics recycling. *See* opt-in organics recycling
volunteer labor, 84–90, 93, 104, 143–44; and data, 93–94; as education, 88; and neoliberalism, 12–14, 113–14, 153; as research method, 23

Waring, George E., 33
waste: building community with, 135–38; markets for, 10–16, 33, 53–56, 72, 114, 160–64; relationships with, 11, 99, 167–69; and stigma, 5–6, 130.

waste (*continued*)
See also food waste; organics recycling
waste labor: at Big Reuse, 88–90; at Earth Matter, 84–88; household, 32, 42n9, 71, 90–92; invisibility of, 4–6, 152–53; at KWL, 138; and utopia, 157, 175; valuing, 93, 119, 135, 146–55, 162, 164. *See also* volunteer labor
Waste Management, Inc., 7, 21, 61, 64, 65, 67, 81
water quality monitoring, 105
WeCare Organics, 21, 61, 64
Westerleigh pilot, 56, 61
Western Queens Compost Initiative, 102, 171
West Harlem Environmental Action Coalition, 53n8
West Oakland Farmers Market, 154
"Whose Streets?" mural project, 140
windrows, 36, 61, 63, 84–90, *86*, 110, 145, 173
Windsor Terrace, Brooklyn, 57
Wolch, Jennifer, 126
worker co-ops, 138, 142
worker safety, 156
worm bins, 156, 156n17
Wright, Beverly Hendrix, 130
Wyly, Elvin, 123

youth employment, 117, 118, 146–47. *See also* BK ROT
Youth Leaders Program, 142

zero waste: community-based reuse enterprises, 18; and environmental justice, 131; and New York City, 2, 49, 58, 71, 176; products, 12. *See also* "0X30"